What people a ying about

Resetting (

A Chicken Can't Duck Egg

Read and act on the wisdom of this book now. In another decade it will be too late.
Dr Susan George, President and Chairman of the Supervisory Board, Transnational Institute.

This book offers a trenchant analysis of the omni-crises that the world now faces with climate change, biodiversity loss, Covid pandemic and financial collapse. It makes compelling arguments for all of us to act now, and change all the systems that underlie human cultures and economies. A must read for every thinking person.
Dr Paul Shrivastava, Director, Sustainability Institute and Professor of Management, The Pennsylvania State University

This cry from the heart from two top policy insiders is a must read! The authors correctly support all the necessary grassroots uprisings for our common planetary future, especially by our children and grandchildren. Global elites are failing. They must give way to servant leaders who grasp humanity's remaining ten-year window to shift lifestyles and cultures toward planetary awareness and restoring its damaged biosphere ,the basis of all life on Earth.
Hazel Henderson, futurist and an economic iconoclast. Author of *The Politics of the Solar Age* (1981) and *Mapping the Global Transition to the Solar Age: From Economism to Earth Systems Science* (2014)

The greatest threat facing humanity is climate change. Whilst we now have solutions, we lack leaders with the ability to implement them. New thinking, a new narrative and community commitment to force unprecedented change is vital. The pandemic, climate change on fast-forward, shows us that the "impossible" is actually possible when needs must. It is humanity's greatest opportunity to turn a fractious and fragmenting world into a genuinely sustainable future. We cannot waste it. The Maxtons clearly set out the challenge, the urgency and pathways to that future. Let's get behind it.
Ian Dunlop, Chairman, The Australian National Wildlife Collection Foundation

This book is bold, dramatic, and visionary. It brilliantly explains the many crises facing humanity and shows why Covid-19 provides a unique opportunity to build a better future. I hope humanity is up to the challenge.
Jorgen Randers, Professor emeritus of climate strategy, BI Norwegian Business School, Oslo. Co-author *The Limits to Growth* (1972) and author *2052 – A Global Forecast for the next Forty Years* (2012)

Will our societies go back to "normal" after the coronavirus pandemic, or will we turn crisis into opportunity and finally get serious about climate change? This is an engaging and unsettling little book that makes a powerful case for urgent and fundamental social change. You may not agree with everything the authors say, but they'll make you think hard, and they'll show you how to get to work with others on tackling the climate crisis.
Mark B. Brown, Professor of Political Science, California State University, Sacramento, USA

This book is an important intervention at a point where we face a historic choice: either to continue an economic system that is driving us to the precipice, or to create new social structures that allow both ecosystems and human societies to survive and to thrive.
Fabian Scheidler, Editor Kontext TV, author, *The end of the mega-machine* Zero Books, 2020

Bernice and Graeme rightly argue that Covid-19 is a chance to accelerate the needed transformation. They point out that this requires escaping the colonial mindset that has made the urban elite blind to their embeddedness in nature. How can we ignore our host planet, and be blind to the destruction of the essential life-support it provides? Covid-19 helps us recognize that we are all one biology and that protecting ourselves is not just good for us, but it is the most effective way to protect others. Indeed, Covid-19 may be the last exit ramp on the highway to disaster. Take Bernice and Graeme's advice.
Dr Mathis Wackernagel, Founder and President, Global Footprint Network

Humanity's social, intellectual, religious, and cultural illness is destroying Planet Earth and depriving our children of a future. This brilliantly constructed, well-documented, no-punches-pulled argument for personal and global change entails tossing out old economic models (no more money), frankly facing the immediacy of the climate crisis, refocusing society from the individual to the community, admitting the weaknesses of unbridled capitalism, ending colonialism and other forms of exploitation, focusing on conservation, and embracing veganism. Buy ten much-needed copies, one for yourself and the rest for forward-thinking, committed, tough-minded, and effective friends who really want to save the world.
Daoist Monk Yun Rou

While many environmentalists are flummoxed over how to stop global warming because they can't see beyond tried and failed carbon taxes, Graeme and Bernice Maxton are scintillatingly clear: "We face an unprecedented crisis ... There is no market-based solution to these problems." There is no tech solution either. "The only way to avoid disaster is if almost everyone cuts their GHG emissions by at least 7% a year. . . This means 20% fewer cars in three years, 20% fewer planes, 20% fewer coal-fired powered stations, and 20% fewer ships. . . By 2040 they need to be zero. Societies also need to change the way they grow food, and stop all deforestation. Everything needs to change." We need a political and social revolution to create political and social systems that "function for the common good." The Maxtons contend that the only alternative to collective eco-suicide is a radical democratization of society to insure that the sacrifices we must make to save ourselves are shared equally and the benefits — a better mode of living, less work, better work — are enjoyed by all. I couldn't agree more. Young people, especially, should read this book!
Richard Smith, economic historian and author, *Green Capitalism, The God that Failed* (2016) and *China's Engine of Environmental Collapse* (Pluto Press, July 2020).

At a time when governments serve mainly the very rich, life- and planet-saving scientific knowledge is useless, unless a way is found to bring such knowledge straight to the people, penetrating their brains, hearts, and consciousness, triggering action. This book does just that, ever so effortlessly.
Chien-Yi Lu, Institute of European and American Studies, Academia Sinica and author, *Surviving Democracy—Mitigating Climate Change in a Neoliberalized World*

The Ancient Greeks had a word for this – a time of chaos, change and yet also opportunity. They called it Kairos. This book is

Kairos incarnate, passionate about the realities; clear about the changes we need; funny when it needs to be and pioneering in that it takes us from chaos to opportunity. Time for Kairos and this is the handbook.
Martin Palmer, Secretary General of the Alliance of Religions and Conservation (ARC), CEO of FaithInvest

This book pulls no punches. No misplaced faith in human technological ingenuity, or renewable solutions that magically occur without massive extraction of raw materials. We need a rapid and far-reaching transformation without further delay. Covid-19 shows us that governments can find money when they want to and carbon emissions can be reduced drastically — but this level of reduction must continue every year. We have to forget about "getting back to normal" and abandon the pursuit of economic growth fueled by endless material consumption. We must aim for balance with Nature instead. This book maps a communication pathway: get the facts straight, bust the myths, and speak out to all around us whether family, friends, school or job. Then target the climate criminals. A clarion call.
Dr Kerryn Higgs, Associate, University of Tasmania, author of *Collision Course: Endless Growth on a Finite Planet* (MIT Press 2014)

Reading this book felt like having some smart, funny, and passionate people round for a dinner party where they explain things in a way that gets the point across while also being witty and engaging. Covid-19 has changed the world in radical ways - and that's exactly the point. The authors urge humanity to take this opportunity, where we've realized we can make huge shifts to the way we live, and harness it to fight climate change. If we don't, the consequences will make 2020 seem like a nice year. Opening with a story about eating breakfast with the Pope and sprinkled with references from Malcom X (the titular epigram) to Monty Python, *A Chicken Can't Lay a Duck Egg* expresses the

urgency of our planet's situation while also suggesting new ways forward.
Dr. Nazanin Zadeh-Cummings, Associate Director of Research, Centre for Humanitarian Leadership, Melbourne, Australia

This book clearly outlines stark truths about impacts of climate change that political leaders of all stripes are unwilling to deliver because the news about what we face is so unsettling. We also read how the best responses to Covid-19 show the scale of what is needed to change direction on climate and that such changes are possible. Highly recommended!
David Berry, Director, Sustainable & Resilient Resources Roundtable

Resetting Our Future

A Chicken Can't Lay a Duck Egg

How Covid-19 can solve the climate crisis

RESETTING OUR FUTURE

A Chicken Can't Lay a Duck Egg

How Covid-19 can solve the climate crisis

Bernice Maxton-Lee
Former Director, Jane Goodall Institute, Singapore

Graeme Maxton
Climate activist, economist, and best-selling author

CHANGEMAKERS
BOOKS

Winchester, UK
Washington, USA

JOHN HUNT PUBLISHING

First published by Changemakers Books, 2021
Changemakers Books is an imprint of John Hunt Publishing Ltd., No. 3 East Street,
Alresford, Hampshire SO24 9EE, UK
office@jhpbooks.com
www.johnhuntpublishing.com
www.changemakers-books.com

For distributor details and how to order please visit the 'Ordering' section on our website.

ISBN: 978 1 78904 761 5
978 1 78904 762 2 (ebook)
Library of Congress Control Number: 2020945780

A CIP catalogue record for this book is available from the British Library.

Design: Stuart Davies

Printed and bound by CPI Group (UK) Ltd, Croydon, CR0 4YY
Printed in North America by CPI GPS partners

Contents

The *Resetting Our Future* Series

At this critical moment of history, with a pandemic raging, we have the rare opportunity for a Great Reset – to choose a different future. This series provides a platform for pragmatic thought leaders to share their vision for change based on their deep expertise. For communities and nations struggling to cope with the crisis, these books will provide a burst of hope and energy to help us take the first difficult steps towards a better future.
– Tim Ward, publisher, Changemakers Books

What if Solving the Climate Crisis Is Simple?
Tom Bowman, President of Bowman Change, Inc., and Writing Team Lead for the U.S. ACE National Strategic Planning Framework

Zero Waste Living, the 80/20 Way
The Busy Person's Guide to a Lighter Footprint
Stephanie Miller, Founder of Zero Waste in DC, and former Director, IFC Climate Business Department.

A Chicken Can't Lay a Duck Egg
How COVID-19 can Solve the Climate Crisis
Graeme Maxton, (former Secretary-General of the Club of Rome), and Bernice Maxton-Lee (former Director, Jane Goodall Institute)

A Global Playbook for the Next Pandemic
Anne Kabagambe, World Bank Executive Director

We Should have Seen it Coming
How Foresight can Prepare us for the Next Crisis
Bart Édes, North American Representative, Asian Development Bank

Impact ED

A Roadmap for Restoring Jobs & Rebuilding the Economy

Rebecca Corbin (President, National Association of Community College Entrepreneurship), Andrew Gold and Mary-Beth Kerly (both business faculty, Hillsborough Community College).

Power Switch

How Activists can win the Fight Against Extreme Inequality

Paul O'Brien, VP, Policy and Advocacy, Oxfam America

Creating a Paradigm Shift to Achieve the Global SDGs

A SMART Futures Mindset for a Sustainable World.

Dr. Claire Nelson, Chief Visionary Officer and Lead Futurist, The Futures Forum

Reconstructing Blackness

Rev. Charles Howard, Chaplin, University of Pennsylvania, Philadelphia.

Cut Super Climate Pollutants, Now!

The Ozone Treaty's Urgent Lessons for Speeding up Climate Action

Alan Miller (former World Bank representative for global climate negotiations) and Durwood Zaelke, (President, The Institute for Governance & Sustainable Development, and co-director, The Program on Governance for Sustainable Development at UC Santa Barbara)

www.ResettingOurFuture.com

This book is dedicated to the remarkable people
of Taiwan.
We thank them for giving us a home, and treating us with
decency and care when our hearts were sore, and our
spirits were worn.
When Covid-19 appeared, Dr. Chen Shih-chung and his
colleagues in the Ministry of Health and Welfare, kept
us safe.
Thank you "Uncle Chen."

Foreword

by Thomas Lovejoy

The Pandemic has changed our world. Lives have been lost. Livelihoods as well. Far too many face urgent problems of health and economic security, but almost all of us are reinventing our lives in one way or another. Meeting the immediate needs of the less fortunate is obviously a priority, and a big one. But beyond those compassionate imperatives, there is also tremendous opportunity for what some people are calling a "Great Reset." This series of books, *Resetting Our Future,* is designed to provide pragmatic visionary ideas and stimulate a fundamental rethink of the future of humanity, nature and the economy.

I find myself thinking about my parents, who had lived through the Second World War and the Great Depression, and am still impressed by the sense of frugality they had attained. When packages arrived in the mail, my father would save the paper and string; he did it so systematically I don't recall our ever having to buy string. Our diets were more careful: whether it could be afforded or not, beef was restricted to once a week. When aluminum foil – the great boon to the kitchen – appeared, we used and washed it repeatedly until it fell apart. Bottles, whether coca cola or milk, were recycled.

Waste was consciously avoided. My childhood task was to put out the trash; what goes out of my backdoor today is an unnecessary multiple of that. At least some of it now goes to recycling but a lot more should surely be possible.

There was also a widespread sense of service to a larger community. Military service was required of all. But there was also the Civilian Conservation Corps, which had provided jobs and repaired the ecological destruction that had generated the Dust Bowl. The Kennedy administration introduced the Peace

Corps and the President's phrase "Ask not what your country can do for you but what you can do for your country" still resonates in our minds.

There had been antecedents, but in the 1970s there was a global awakening about a growing environmental crisis. In 1972, The United Nations held its first conference on the environment at Stockholm. Most of the modern US institutions and laws about environment were established under moderate Republican administrations (Nixon and Ford). Environment was seen not just as appealing to "greenies" but also as a thoughtful conservative's issue. The largest meeting of Heads of State in history, The Earth Summit, took place in Rio de Janeiro in 1992 and three international conventions — climate change, biodiversity (on which I was consulted) and desertification — came into existence.

But three things changed. First, there now are three times as many people alive today as when I was born and each new person deserves a minimum quality of life. Second, the sense of frugality was succeeded by a growing appetite for affluence and an overall attitude of entitlement. And third, conservative political advisors found advantage in demonizing the environment as comity vanished from the political dialogue.

Insufficient progress has brought humanity and the environment to a crisis state. The CO2 level in the atmosphere at 415 ppm (parts per million) is way beyond a non-disruptive level around 350 ppm. (The pre-industrial level was 280 ppm.)

Human impacts on nature and biodiversity are not just confined to climate change. Those impacts will not produce just a long slide of continuous degradation. The Pandemic is a direct result of intrusion upon, and destruction of, nature as well as wild-animal trade and markets. The scientific body of the UN Convention on Biological Diversity warned in 2020 that we could lose a million species unless there are major changes in human interactions with nature.

We still can turn those situations around. Ecosystem restoration at scale could pull carbon back out of the atmosphere for a soft landing at 1.5 degrees of warming (at 350 ppm), hand in hand with a rapid halt in production and use of fossil fuels. The Amazon tipping point where its hydrological cycle would fail to provide enough rain to maintain the forest in southern and eastern Amazonia can be solved with major reforestation. The oceans' biology is struggling with increasing acidity, warming and ubiquitous pollution with plastics: addressing climate change can lower the first two and efforts to remove plastics from our waste stream can improve the latter.

Indisputably, we need a major reset in our economies, what we produce, and what we consume. We exist on an amazing living planet, with a biological profusion that can provide humanity a cornucopia of benefits—and more that science has yet to reveal—and all of it is automatically recyclable because nature is very good at that. Scientists have determined that we can, in fact, feed all the people on the planet, and the couple billion more who may come, by a combination of selective improvements of productivity, eliminating food waste and altering our diets (which our doctors have been advising us to do anyway).

The *Resetting Our Future* series is intended to help people think about various ways of economic and social rebuilding that will support humanity for the long term. There is no single way to do this and there is plenty of room for creativity in the process, but nature with its capacity for recovery and for recycling can provide us with much inspiration, including ways beyond our current ability to imagine.

Ecosystems do recover from shocks, but the bigger the shock, the more complicated recovery can be. At the end of the Cretaceous period (66 million years ago) a gigantic meteor slammed into the Caribbean near the Yucatan and threw up so much dust and debris into the atmosphere that much of biodiversity perished. It was *sayonara* for the dinosaurs; their only surviving close relatives were precursors to modern day birds. It certainly was not a good time for life on Earth.

The clear lesson of the pandemic is that it makes no sense to generate a global crisis and then hope for a miracle. We are lucky to have the pandemic help us reset our relation to the Living Planet as a whole. We already have building blocks like the United Nations Sustainable Development Goals and various environmental Conventions to help us think through more effective goals and targets. The imperative is to rebuild with humility and imagination, while always conscious of the health of the living planet on which we have the joy and privilege to exist.

Dr. Thomas E. Lovejoy is Professor of Environmental Science and Policy at George Mason University and a Senior Fellow at the United Nations Foundation. A world-renowned conservation biologist, Dr. Lovejoy introduced the term "biological diversity" to the scientific community.

Introduction

It is disconcerting to come out of your room in the Papal Guesthouse in the Vatican and bump into the Pope himself, alone, in his full white robes, by the stairs. What should you say, exactly, and in which language? At least the experience made breakfast at the table next to his Holiness ten minutes later seem almost normal.

Other strange experiences have peppered our lives. It felt cool to approach No.10 Downing Street in London, give your name to the police officer, and be swiftly welcomed inside. Even more, to receive a friendly wave of recognition from the Prime Minister, before chatting with the head of the British government's policy unit, and then travel to Tehran a few days later for a meeting with Iran's Finance and Development Minister. Lunch, alone with Carlos Ghosn in Sao Paolo, had a surreal quality too, as did chairing a meeting of the world's Central Bank Governors in Bali with a tank sitting on the lawn to protect us. Sitting in the cockpit next to Ratan Tata, as he piloted his private jet across India, felt good but also rather strange, especially when it came hours after shaking the dying hands of leprosy patients. It was a special thrill to address the United Nations in Geneva in 2018 on the topic of climate change.

Yet, it was also disappointing to realize, after all this, that the world does not function as it should.

For many years we worked to promote radical social reform. We didn't do this for some moral or religious purpose. We did it because there is no other way. If societies do not change, they will collapse. Just over a year ago we gave up. We were exhausted, heart-sore, and burned out.

Together, we had worked as environmentalists for more than 30 years, though we did not start there. We began in the glitzy, high-risk, high-reward world of banking, strategy consulting, and economics. Over the years, we began to ask some difficult questions but found the answers, or the lack of them, troubling.

Before the financial crisis of 2008 we worked in China and South East Asia and we found deep inconsistencies between what we saw with our eyes and the narrative of progress. Sure, these countries had lots of economic growth. But the lives of most people were very harsh, and their days endless. While millions had been lifted out of poverty, they had also been taken out of a sustainable way of living, dumped in the money economy and forced to work 15-hour days. And the environmental impact of this development was horrendous. Plastic waste, stagnant oil-coated rivers, stinking piles of burning, toxic rubbish, and burgeoning mega-cities had replaced the lush green paddies we had seen 20 years before. So we began to ask ourselves: is this what economic growth achieves? Our peers and colleagues, including those in *The Economist* magazine where we worked, could not answer our questions.

We worried that we were part of a machine which was the cause of misery, not progress.

That was the first time we stepped back. Our questions took us from Asia to Europe, from economics to academia and the environmental movement. We retrained, learning mid-career about climate chemistry, international relations, and environmental law.

Hoping that we could do something useful with this knowledge, we landed back in Asia some years later, this time in Singapore. There, we experienced firsthand the choking smoke billowing over from the rainforests, as they burnt in Indonesia. Again, we asked "why?" This time we were met with disinterest, and hostility, especially by those in power. Even our colleagues at the Jane Goodall Institute preferred not to discuss why the habitat of rare and ancient primates was going up in smoke.

So we dug deeper, invested more of ourselves and more of our energy, to learn more, lobby harder, and push for change.

We wrote books; one of us got a PhD on why efforts to stop deforestation aren't working, while the other rose to the very top of one of the world's great environmental think-tanks, as Secretary General of the Club of Rome. We addressed the United Nations, met the Pope, engaged with influencers, and convened with Presidents, Prime Ministers, and Central Bank Governors around the world. From there, surely, we would finally understand the problem.

And so we did. We could see that decision-makers, heads of state, and the bosses of big corporations were all too obsessed with making money to listen to explanations about the destruction of nature and society, or what would happen without change.

We felt like we had no more to give. The time remaining to prevent collapse was too short, and not enough people were listening. So we left our home in Switzerland with plans to spend a year in Taiwan, far away from everything familiar. Shortly after, the coronavirus began its deadly advance across the world.

As the crisis grew, we saw a final chance for change. It boils down to this:

For decades, most of the economically dominant governments of the world have failed in their obligations to the majority of their citizens. Anticipating their flailing, incompetent response to Covid-19 was easy. It followed a well-established pattern.

1. It is not possible to reform the economic, political, and social systems of these countries to make them function in the interests of most people. A chicken cannot lay a duck egg.
2. In the coming decade, humanity will be overwhelmed by a series of self-inflicted, interconnected ecological crises. The impact of climate change will completely overshadow the effects of Covid-19.
3. There is no market-based solution to these problems. Solar

panels, Teslas, and some as-yet-undefined technology will not save us.

4. Only a people's movement and structural reform will work.
5. Covid-19 makes this possible.

The struggle that lies ahead is necessary because corrupt, rich-world governments have been holding out, pushing back against the need for change through denial, incompetence, and an inability to tell the difference between fact and fiction. Those elected to power, those who finance them, and those who control the world's big corporations have chosen instead to protect their own uneven material wealth and power. They have continued to colonize the world's resources and people, controlling and exploiting them for short-term gain.

With mounting ecological devastation, these governments, and the powerful, are failing in their basic obligations. Hundreds of millions of people are being denied a decent standard of living. Without change, future generations will inherit a broken, shattered world.

The time for these governments to act is not when the consequences of their negligence and folly have become overwhelming, and are unfixable. It is now, and every day from now for the next decade. That is the only time societies have left to introduce radical changes to the way people live, think, and dream.

After then it will be too late. There may be an even bigger push for change, and perhaps angry revolution but, by that stage, it will be in vain.

The change that is needed should be led by the young, because they have the most to lose. Those who are in their 20s and 30s will spend most of their lives in a rapidly-warming world, with all the chaos and violence this will bring. Young change-makers are also less likely to have been brainwashed by the economic propaganda that is killing the planet. Those who have not yet been infected with this dangerous virus, which destroys and divides, will be better able to build a different, less selfish future for humanity.

We, the people, have an obligation to hold in front of us a burning torch to illuminate the purpose of our struggle. From this day, and every day for the next decade, our goal is still achievable, though only just.

Do not leave it until that struggle becomes pointless. Let us all make each day count.

September 2020.

Part 1

Trouble bubbling

Chapter 1

Never waste a crisis

At the core of the struggle that lies ahead is a decision on how people choose to live. Should they continue to make decisions, as they have done for so long, based on humanity's selfish, short-term animal instincts? Or can they finally learn to favor the reflective and spiritual side of the human character?

They have very little time to choose.

As we write, the social foundations of many countries are crumbling, hundreds of millions of people have lost their jobs and thousands are dying every day. All this, the result of a virus which no one had heard of a year ago.

As well as being the cause of so much misery, the coronavirus has brought humanity something that is almost beyond value. Woven within the economic destruction and death is the greatest opportunity for radical social and economic reform that societies have seen in decades. What the future holds, and where the opportunities lie, is the subject of this book.

Most of this book is about the long-term outlook, beyond the next few years. But what of the time before? From our virus hideaway in the late summer of 2020, three possible short-term pathways lay ahead:

1. After months of governments fumbling, especially in the US, Brazil, Spain, and the UK, with rising infections and businesses struggling to survive, the first possible future we could see was for more of the same. The turmoil would simply continue. There would be weeks of optimism, when infections fell and the economy recovered. But these would be broken by the discovery of new virus clusters and further enforced shutdowns. We saw this future a bit like a Salvador Dali painting, a recurring nightmare punctuated by surreal happy moments.

2. The second, more positive, option (though not for us) we could see was a recovery. Like the movie, societies would go "back to the future" with their economies slowly returning to how they were before the virus struck. We thought this might happen if a low-cost, easy-to-administer vaccine were developed. Or it could happen because the virus simply blows itself out, mutating into a less infectious form, as happened at the end of the great plague of London in 1665.

This second future was pretty much guaranteed, at least in some form. Given the vast amount of money that governments around the world had invested in developing a vaccine, the fact that many were competing with each other, the huge profits promised to those that succeeded, and the wider economic advantages that would come to those who controlled the spread of the virus first, it seemed a sure bet that several Covid-19 vaccines would be launched quickly, and perhaps even before our book was published.

But we could also see a problem. Any speedily launched vaccine would have been developed much faster than usual. Given the high stakes, and the fact that the multinational pharmaceutical companies had been given a legal liability waiver, so it did not matter if their vaccines actually worked or had any nasty side-effects, there was a risk that this best-case scenario would be characterized by a series of false dawns.

And, even if a safe, low-cost vaccine were developed successfully, it would take several years to administer to nearly 8 billion people, as the virus continued to create havoc. Clusters of infections would still pop up unexpectedly, requiring further unpredictable restrictions.

So even if this positive future came to pass, and even if there was a safe vaccine, life would probably still be hard for most people, and not very different from the first future we imagined.

3. The third possible future we could see was rebirth. This could happen because societies gradually came to realize that

they had hit a dead-end and could see no chance to return to their past normality. It could happen if it proved impossible for scientists to develop a safe, reliable, long-lasting, and effective vaccine. Or it might happen because a more deadly mutation of the virus emerged, leading to a widespread semi-permanent lockdown.

It was the third possible future which was of the greatest interest to us. While a prolonged crisis could lead to conflict, as the peoples of the world sought to blame others for their plight, the descent into this dark place also offered the chance for societies to make a radical overhaul to the way people lived, thought, and worked. If successful, this could take societies to a much better place, with a renaissance in human behavior.

It is during a crisis that radical change is often possible. We saw Covid-19 as a doorway, a portal to a different future, where humanity could address its bigger, long-term challenges.

Why did we think that change on such a vast scale was needed? That is the subject of the first part of the book.

Humanity screwed up

A radical change in direction is necessary because humanity faces a series of crippling environmental crises during the next decade which, without change, will make the impact of Covid-19 seem like a picnic. The first of these crises is directly related to the emergence of the virus.

As people encroach further into the territory of other animals, their rate of extinction is accelerating, and the number of diseases jumping from other species to people is growing. In areas where natural systems have been badly degraded, the number of animals which host these "zoonotic" diseases, such as bats and rats, is 250% higher than usual[1] while the proportion carrying the pathogens is 70% greater. SARS, Zika, HIV, MERS, Ebola, and Covid-19, as well as many other deadly diseases, are all a result of humanity's disregard for nature.

Covid-19 has caused the biggest global pandemic for more than a century, but it will be the first of many unless there is change.

The second looming ecological crisis is rising pollution. It is one of the main causes of species die off. The steady accumulation of micro- and nano-plastics in the world's oceans and rivers has been described as the "number one threat" to humankind[2] while "nine out of ten people breathe polluted air."[3] It kills 7 million people a year, with respiratory diseases the third biggest cause of human mortality.

Serious environmental problems are also being caused by mining, energy production, and urbanization. Dams built to generate electricity have caused earthquakes. The extraction of oil, gas, and coal from deep underground is leaving unstable, polluted landscapes. The world's rainforests are being destroyed, while increasing soil erosion will limit the ability of future generations to feed themselves.

These environmental problems are urgent because humanity's destructive impact has grown so quickly. The number of people has more than doubled in 60 years and is almost five times greater than a century ago. Even after accounting for the deaths caused by Covid-19, the human population is still rising by 80 million a year. That is another billion every 12 years; a billion more people needing food, water, housing, clothing, and waste management. With the push for more economic output requiring ever more energy, land, and raw materials, as well as rising levels of urbanization, the accumulated impact of humanity's activities on nature has become overwhelming. This is especially so when it comes to climate change, which is by far the most serious looming environmental catastrophe of all.

Chapter 2

Enlightening and frightening

It's easy to get confused about climate change. The endless headlines can be as numbing as the endless inter-governmental meetings. The problem is presented as urgent and yet people are also told that the most serious consequences are decades away. You probably know too, that there is a lot of misinformation out there, with fossil fuel firms sowing seeds of doubt about the science or denying there is a problem. So, what's the truth? Buckle up, because the following pages may enlighten (and frighten) you more than you imagine. They will help you understand why everything that societies are currently doing in response to climate change will fail. All those investments in wind farms, solar energy, electric cars, and recycling will, on their own, have almost no useful impact.

Scientists have known for decades that the planet's average temperature has been rising, and that the pace is accelerating. They know too that what is happening is not natural. It is happening too quickly to be part of any atmospheric cycle, and there has been no natural event, like a volcanic eruption, to otherwise explain it.

The warming is mostly down to the way humans produce energy and food. This creates gases, known as greenhouse gases, in greater quantities than nature can absorb. Most of the excess gases remain in the atmosphere where they trap some of the sun's heat. This is the greenhouse effect. Because of this, the average global surface temperature is 1.1ºC higher than it was 200 years ago. While that might seem a small increase, it is now higher than at any time in the last 3 million years.

The increase is already causing a lot of problems. Other species are dying, especially in the world's oceans, which are warming faster than the land. Mountains are crumbling as the ice that holds them together melts. Some crop yields are declining. Glaciers are disappearing and forests are dying. Storms are becoming more frequent, the number of wildfires is growing, and droughts are becoming more prolonged. The permafrost in Canada and Siberia is also melting, releasing the gases that were trapped beneath for tens of thousands of years. It is also releasing deadly anthrax spores which have infected reindeer herds and the people who live nearby.

The melting permafrost and decaying forests are adding to the warming problem because the gases they release are also greenhouse gases. As the world's glaciers and polar ice melts, less heat is reflected into space. This is making the planet hotter too.

If the concentration of greenhouse gases continues to rise at the current rate, the world will reach a catastrophic tipping-point in the mid-2030s. If this is breached, the warming will be impossible to control and by the middle of the century the average temperature will have reached its highest level in 10 million years. By 2100, the Earth will be on track to become as hot as it was 45 million years ago.

Think about how the planet looked 45 million years ago. There was no ice and very few living creatures other than fish, lizards, and insects. We are not saying that this is how the world will look in 2100. It will take centuries for the rise in temperature to return the earth to how it once was. Even so, if this chain-reaction begins, many parts of the planet will be uninhabitable by the second half of this century. This, by the way, is also what will happen if all of the conditions of the 2015 Paris Climate Accord are met. What has been agreed by governments in response to climate change so far will not avoid this catastrophe, nor delay it by one second.

Of course, many people understand that there is a climate problem, though they may not fully understand exactly how large and urgent it really is. But there are already a huge number of people working hard to make their societies more sustainable. They buy electric cars, recycle their waste, avoid plastic packaging, take fewer flights, and sell their fossil energy investments. Green groups around the world are working hard too, pushing governments and businesses to invest in renewable energy.

Hard as this might be to understand, none of these activities will achieve anything like the change that is needed. Despite their good intentions, activities on this scale are too small to stop what is happening to the planet.

The scale of the climate problem is so big that, even if hundreds of millions of people lived 100% sustainably, and created absolutely no greenhouse gases, it would not be enough to avoid the tipping point.

The main greenhouse gas is carbon dioxide (CO2). Before the industrial revolution, in the early nineteenth century, the concentration of CO2 in the atmosphere was 280 ppm (parts per million). It had been like that for many hundreds of thousands of years. After people began burning fossil fuels, however, the concentration of CO2 began to rise.

In 2020 it was almost 50% higher, at 416 ppm, and growing by 3 ppm a year exponentially. The tipping point that societies have to avoid, when the chain-reaction starts, is when the concentration reaches 450 ppm. This is the level which will gradually take the planet back more than 45 million years, to the last time the CO2 concentration was 450 ppm.

That is in less than 15 years.[4]

If humanity is foolish enough to let this happen, the great forests around the world will die and the ice at the poles will melt even faster. Mountain glaciers and coral reefs will disappear. There will be a steady and uncontrollable rise in global temperatures over centuries. Most of the planet will eventually become uninhabitable, killing off up to 95% of the human population.[5]

How can societies avoid this?

When people learn about what is happening, the first question they ask is "what can I do?" We will answer this question in detail at the end. For now, though, let's assume you live in the rich world. What would happen if you found a way to live 100% emissions free? We are not talking about buying a Tesla and taking fewer flights. We are talking about living off-grid, selling your car, turning vegan, and doing everything possible to cut your contribution to the atmospheric pollution to zero. Let's say you start tomorrow. What impact will that have over a decade?

Your sacrifices would delay the start of the chain-reaction by a fifth of a second.

Even if *everyone* in America – all 330 million people – had some sort of epiphany tomorrow and lived without generating any damaging emissions for the next decade, it would only delay the onset of disaster by two years. This is because the US is only responsible for 15% of emissions (a lot, when it has only 4% of the global population). But if those responsible for the other 85% continue as now, America's efforts would not be enough to stop what is happening. They would only delay it a short time.

And, even this is before taking account of the emissions from the wildfires in California, Australia, and elsewhere, which are pumping vast quantities of CO2 into the atmosphere. Nor does it count the powerful nitrous oxide emissions from fertilizer runoff, or the rising methane emissions from abandoned coal mines, and garbage tips. Although the volume of these gases is much smaller than CO2, they have a much greater warming effect – up to 300x more.

In the face of such an enormous problem, acting alone can't have any useful impact, just as converting every car to an electric vehicle or closing every coal-fired power station in Europe and North America will not be anything like enough either. Unless the response includes China, Russia, Japan, India, and Australia, as well as North America and Europe, humanity cannot slow the pace of warming quickly enough.

Humanity is battling the laws of chemistry on a planetary scale.

It is also important to understand that humanity cannot stop climate change, certainly in any time frame most people can grasp. All societies can do is make sure the warming doesn't get out of control. Even then, because of lags in the atmospheric system, and the release of gases from wildfires, mines, garbage dumps, soils, and beneath the Arctic permafrost, global temperatures will continue to rise for many years to come.

What to do?

The only way to avoid disaster is if almost everyone *cuts* their GHG emissions by at least 7% a year.[6] In practical terms, this means 20% fewer cars in three years, as well as 20% fewer planes, 20% fewer coal-fired powered stations, and 20% fewer ships. In the following three years there needs to be another 20% reduction. And the longer societies take to begin this process, the steeper the cuts have to be. To work, greenhouse gas emissions must be at least 60% lower in 2030 compared to today.[7] By 2040 they need to be zero. Societies also need to change the way they grow food, and stop all deforestation. They will also need to build thousands of carbon capture and storage plants, and run them full-blast for decades, to bring the CO2 concentration in the atmosphere back down to safer levels. Even then, even having done all this, humanity's chance of avoiding that chain-reaction will be little better than 50:50.

And, to be clear, when we talk about cutting emissions to zero, we do not mean "net-zero" as some fossil fuel companies, airlines, and governments suggest is okay. Trying to offset emissions in some way, such as planting trees, which take decades to grow, will not have anything like enough impact on what is happening, just as taking exercise does not offset the effects of a 20-a-day cigarette habit when someone has been diagnosed with lung cancer.

When someone asks us what they can do, our answer is simple: "If you act alone, there is nothing you can do. Everyone must change." You can only set an example for others to follow. And part of your plan must be to persuade others to change too. More on this later.

For now, the message is this: to avoid a catastrophe, almost everyone needs to change the way they live, and they need to do this urgently, whether they want to or not. Polluting businesses – fossil fuel firms and cement companies – have to be shuttered as fast as possible, most flights have to be permanently canceled, and vehicle use has to be hugely curtailed. Most electric vehicles need to be broken up and recycled too, if they recharge their batteries with electricity from coal, gas or oil-fired power plants. Put simply: the way most of us live has to be dismantled, regardless of the economic and social consequences.

Without a change on this scale there is no point in doing anything at all. You and your family turning vegan will not save humanity now. The atmospheric changes that have been set in motion are way beyond that.

Whatever happens, there *will* be a transition to another system of human development, whether it is the one we propose in this book, or another, more confrontational and chaotic one. At some point soon, the failures in the current system, the impact of climate change, and the planet's other many environmental troubles will come together and force change.

The position humanity has created for itself is grave, but it is not yet hopeless. This is where Covid-19 comes in useful.

Chapter 3

What stops societies changing?

Covid-19 has shown that the transformation needed *is* actually possible. It is possible to cut emissions, shut the airline industry, reduce vehicle use, and support people financially. Some countries have done this better than others and the economic cost has been horrendous, of course. The pressure to return to "normal," and put the economic health of the country before the health of people, has been very large too, and few would want to repeat their experience of lockdown. But the virus has still shown that change on the scale needed is possible.

When it comes to dealing with the climate challenge, the changes would obviously need to be larger and permanent. It needs need a structural transformation. Until Covid-19, however, there was a widespread belief that the radical changes and the drastic cuts in emissions had to be financially attractive. Covid-19 has shown that this is wrong.

The problem is that the fossil fuel narrative remains as before. People think that societies can make a gradual transition to a fossil-free world over several decades. The coal, oil, and gas companies believe it will take longer. They claim that 70% of the energy used in 2040 will be fossil-based and expect to prosper into the second half of the century. Without an alternative source of energy that works as well, and at similar cost, a faster transition is thought to be impossible. So societies continue to burn fossil fuel, hoping that the price of renewable energy will fall so investment can rise.

In effect, however, societies are choosing to damage the atmosphere, rather than change. They don't want to swap to an energy source which is more expensive and less convenient, or stop living in a way that appears normal.

This means that at the center of the climate challenge lies a social problem. Societies are choosing not to fix it. They are choosing instead to release more gases. It is a mindset problem, not a technological one. It is humanity's beliefs that are the main barrier to change.

Climate change can only be brought under control, and the wider ecological devastation stopped, if there is a change in thinking. Even without a cheap alternative source of energy, societies have to stop generating gases, and choose a less destructive path. That is not thought possible because the majority of people, as well as many in government, have been persuaded to think that it can't happen unless it makes economic sense. Covid-19 has shown that it can. It has shown that output can be cut, travel can be swiftly reduced and people can change the way they live and work overnight. It is not impossible. It has already been done.

The beliefs which lie at the center of this mental blockage concern human purpose, individual freedom, and democracy.

The idolatry of false ideas

For decades, what modern societies have treasured, and so measured, has not been improvements in living standards or advances in human knowledge. It has been economic growth. They have sought to increase their output, or GDP.[8] There was an assumption that this would improve well-being and lower inequality, leading to greater prosperity for all. But that was just an assumption. And, as it turns out, it is mostly wrong. In the last few decades, the living standards of millions of people have not improved, despite lots of economic growth, while inequalities have widened. This is also true in most of the poor world. While millions of people have been lifted out of poverty in the last 30 years, most remain very poor. More than 90% still live on less than $10 a day while the gap between the rich world and the poor world is three times greater today than in 1820.[9]

Why this focus on increasing output?

For nearly 30 years, after the end of the Second World War, most of the rich world enjoyed high rates of economic growth. At the same time, average living standards greatly improved. This allowed economists to claim that the two were linked. Yet the reality is not so simple.

The high rates of economic growth between the late 1940s and the mid-1970s were partly a consequence of war. Countries had to rebuild and, with a low starting point, the rates of growth were inevitably high. At the same time, living standards improved. But this was not only because of rising output. It was greatly the result of government investments in healthcare, infrastructure, and welfare, especially in Europe.

So, the link between economic growth and rising prosperity is not as many economists have claimed. One does not always lead to the other. This will become clearer in the future, as societies relocate cities away from coastal areas because of rising sea levels, armies manage the flow of migrants, and buildings destroyed by wildfires need to be rebuilt. All these activities will generate lots of economic growth, but none will improve living standards.

It is also possible, incidentally, to improve average living standards without economic growth. Governments can print money, and give it to those in need, or they can redistribute wealth by taxing the rich and giving the money to the poor.

Two other economic ideas promoted in the pursuit of increased output have had much more damaging consequences.

The first is that markets should be minimally regulated. This has been presented by economists and business people as simple common sense – something that is obvious and natural. If businesses are free of government red-tape, goes the thinking, there will be faster economic growth.

This idea has become so widespread that it is hard to remember that it has no logical foundation. There is no natural or logical reason why the basic human activity of buying and selling goods and services should be less regulated than any other. Even so, economists have argued that control of this exchange process should be left to the "invisible hand," citing eighteenth-century Scottish historian and philosopher Adam Smith (he was not an economist). These unseen market forces, and the pursuit of the best outcome by each individual, are all that are needed to ensure the best outcome for all, economists say.

This is not what Adam Smith actually said. When Smith talked about the invisible hand, he was not referring to trade, or how markets should be regulated. He was talking about what he believed was the natural desire in people to help others. When it came to trade, Smith believed that too little regulation would lead to businessmen hijacking the economy to maximize profits at the expense of society.

Smith believed that business people need to be regulated because they have a tendency to conspire together against the public, "People of the same trade seldom meet together, even for merriment and diversion, but the conversation ends in a conspiracy against the public, or in some contrivance to raise prices."[10] Attempts by business people to influence regulation should be treated with great caution, he said, as they "generally [have] an interest to deceive and even oppress the public."[11]

Modern-day economists have re-imagined what Smith said to suit their own purposes (we'll explain why later). In reality, there is very little evidence to support the idea that minimally regulated markets, directed by some invisible ethereal force, serve wider society's long-term interests. There is nothing to suggest that this leads to the best outcome for most people.

Instead, the opposite appears to be true: too little regulation promotes environmentally destructive and unethical business behavior, with a focus on short-term profits – activities that are not generally in society's long-term interests. Despite this, the idea that markets should be lightly regulated has become so deeply embedded that almost no one questions it.

The second damaging economic idea is that unintended consequences should be ignored. If families are torn apart by a company relocating, and a community is forced into decline, that is not of any concern to business people. If nature is damaged, polluted, or destroyed in the pursuit of higher economic growth, this too should be ignored. Economists say that people's livelihoods, the wrecking of communities, and the destruction of nature should be viewed as "externalities," or unintended consequences, just as if lung cancer should be seen as an unintended consequence of smoking. The overriding goal of business is to maximize output and increase short-term profits. It is not the responsibility of companies to protect nature or people. That is the responsibility of governments.

This is a spectacular piece of double-think. Governments are told not to regulate, then dumped with the fallout of too little regulation. They are then pilloried for their incompetence in failing to protect people and nature, creating a vicious cycle, which continuously undermines credibility in the state while strengthening the sanctity of the market. It leads to workers being viewed as little more than cogs in a machine, and the natural world being viewed only as a source of raw materials, and so potential wealth.

Promoting these damaging ideas makes sense for those who want to focus on short-term profit maximization, of course. But that is only a small, wealthy elite. This thinking allows them to justify their actions, even if they create an existential problem for everyone. It is this belief system which allows the bosses of the world's fossil fuel companies to continue their destructive activities, without being prosecuted by the rest of society for knowingly upsetting the chemical balance of the planet. The catastrophe these people create is viewed as an externality, so their accountability is ignored.

It is the desire to increase short-term profit without any

thought for the long-term social consequences that allows more airport terminals to be built, more cars to be sold, and more cruise ships to cross the oceans (until the advent of Covid-19) regardless of the high levels of pollution these activities cause. To those who treasure higher earnings, melting ice caps are good news because they offer the chance of new shipping routes, just as destructive wildfires in Australia and California create opportunities for the construction industry.

It is these dangerous beliefs that have led to the difficulties everyone now faces. To achieve ever-more economic growth requires the use of ever-more resources. To extract and refine these resources requires ever-more energy. With more than 80% of that energy still fossil fuel derived, this creates the pollution that changes the world's climate.

The push for economic growth is the direct cause of climate change.

To slow the pace of global warming, societies have to abandon these ideas. Until that happens, as economic historian Richard Smith puts it, "we're all on board the TGV of ravenous and ever-growing plunder and pollution. As our locomotive races toward the cliff of ecological collapse, the only thoughts in the minds of our leaders is how to stoke the locomotive to get us there faster. We're doomed to a collective social suicide – and no amount of tinkering with the market can brake the drive to global ecological collapse."[12]

Instead of promoting growth, societies need to live in balance with nature, regardless of what that means in terms of jobs lost, or businesses closed, in the short-term. Covid-19, dreadful as it is, has shown societies that this transition is actually possible. It has shown that it is possible to shutter entire sectors quickly and pay workers out of state funds. It has shown that it is possible to reduce the volume of greenhouse gas emissions quickly. Of course, the economic consequences – the externalities – of the virus have been catastrophic. There has been enormous social upheaval, with people demonstrating on the streets, and a deterioration in many people's mental health. International political tensions have also grown.

Yet these difficulties have also shown societies what they need to focus on if they are to slow the pace of climate change. They have shown societies how hard it will be to completely shut all the unnecessary, wasteful, and polluting industries. They have shown societies how hard it will be to support hundreds of millions of people financially.

Covid-19 has taught societies how much they need to invest in the transition if they are to do what is necessary. It has shown them that while a radically different path ahead actually exists, it will not be easy to travel.

Before, societies did not really understand what they were up against. They did not understand the consequences of shutting industries or know how hard those who want to maintain the status quo would fight back. Now they do. That is a huge step forward.

Chapter 4

Did you vote for this?

Another mental blockage lies in humanity's current approach to democracy. This is not just because the system elects comedians, reality TV stars, and faded film stars to power, people who are not qualified for the job. Nor is it just because the system has been corrupted by the influence of big business and wealthy individuals in many countries. It is because modern societies have failed to understand what democracy means.

Democracy means rule by the people. It does not mean rule by representatives of the people. That may seem a minor difference but it is not. In ancient Greece, where democracy began, society was not managed by a small number of elected representatives. It was led by a large group of politically engaged citizens. The Greeks did not just tick a ballot paper every few years and then sit back to criticize those they had elected. A large minority actually ran the country.

Modern democracies are different in two other important ways. First, almost anyone can be elected. Second, voting is open to every adult in most countries.

These might seem valuable improvements. Yet they are very large hindrances when it comes to dealing with a challenge as large and complicated as climate change.

Modern democracies frequently elect people who know almost nothing about the issues they are expected to regulate, other than what they read in the newspapers or on social media. This is especially true when it comes to climate change, where many politicians understand very little about the scale and urgency of the problem. This is made worse by the influence of lobby groups and wealthy sponsors, especially in the US. Rather than focusing on the needs of those who elect them and the long-term interests of society, many of today's politicians dedicate their time to promoting policies that are only in the interests of those who fund them – big businesses and the rich.

Second, the modern democratic system makes it almost impossible for governments to introduce the changes needed. Any government which decided to close damaging businesses and limit personal mobility would be voted out of office. To stop the damaging emissions means the majority of voters have to support the transition, even though most do not understand the scale or urgency of the problem. It also requires people to vote for a change which will make most of them worse off in the short-term, at least materially, and very few people will do that.

Radical change is essential if humanity is to avoid a catastrophe, but the democratic requirement for the majority of people to support a transition makes it impossible. It is an almost insurmountable hurdle. By the time enough people understand what is needed, because they have been terrified by the storms, fires, and floods, the critical emissions level will almost certainly have been breached, and there will be no way back.

Related to the democracy problem is current thinking on individualism.

At the start of the Covid-19 crisis we lived in Taiwan and our friends told us they could not understand the response to the pandemic in Europe and the US. They could not see why so many people refused to follow rules designed for the good of everyone. They were bewildered by reports of people refusing to wear face masks or keep a distance from each other even when the virus was raging out of control.

In Asia it was not like that. Everyone observed the rules for the collective good, with the result that life continued mostly as it had before in many places. There was no lockdown where we were because very few people were infected. Within a few months there were no locally transmitted infections in Taiwan at all. The only new cases were in people returning from overseas, who were identified at the airport and put into quarantine. While life in much of Asia continued as before, America and Europe struggled. "Why are Europeans and Americans so selfish?", people asked us.

It is Western thinking on individualism that makes people behave like this. In much of Europe and the US, most people do not see that there is a collective social interest which needs to be protected, because, to quote neoliberal icon Margaret Thatcher, they believe that "there's no such thing as society." This is not how people think in Asia.

This belief in the sovereignty of the individual encourages people to think that they can stop climate change by altering their shopping, travel, and eating habits. It has led them to think that they are personally responsible for the world's environmental problems, which is nonsense. As we will explain, it is multinational corporations and the finance sector which are almost entirely responsible for humanity's environmental problems. They are not held accountable, however, because economists say that these companies are "only" responding to the demands of the market, which is made up of individuals.

Extreme individualism in Western societies has been encouraged by those in power (by which we mean big business and the rich) because it atomizes society. It makes it easier for big corporations and the wealthy to plunder without restraint because there is no one to hold them accountable. Governments have been told not to regulate and individuals acting alone have no power.

Because they have been told that they are individually responsible for their own lives, happiness, successes, and failures, people see little reason to do anything which might benefit others. The only thing that matters is each ego; their bucket lists, shopping aspirations, and Instagram feeds. Each person "deserves" to be special because they "are worth it" for some reason.

There are, of course, many people who work for the common good. Abolitionists, suffragettes, movements for civil rights and Black Lives Matter are all examples of people rising up to create change. But these movements, powerful and just as they are, have nonetheless allowed the system to continue mostly as it is.

More than 150 years after the end of the American Civil War, black lives are still less economically equal, politically-represented, and judicially-protected than white lives. More than a century after suffragettes chained themselves to railings in Britain, women are still paid less than men for the same jobs in nearly every country.[13]

Anti-war protests have been waged for centuries, against the Second Boer War from 1899–1902, the First and Second World Wars, the Vietnam War, and the invasions of Afghanistan in 2001 and Iraq in 2003. In some rare cases these shortened the duration of the war. History shows, however, that wars have continue to be waged, in spite of the protests, with hundreds of thousands of people still dying and displaced in military conflicts every year. According to the World Bank, conflicts have increased sharply since 2010.[14] In the long view, protests against war have achieved very little.

As for the environmental movement, despite the tireless efforts and sacrifices of so many, for so long, nature is worse off, the climate is more screwed up, and more species are dying, than ever before. Movements in rich countries against acid rain and pollution have succeeded only in the use of more technological patches, as well as a mass-exodus of polluting industries to countries where people's voices are muted and environmental standards are lower (and please note: it is *people* who made the decisions to do this). Even the much-touted success of closing the ozone hole is not quite what it seems. The hole is smaller than it was in 1982 but it is still very large and will remain a problem for decades to come. There are also concerns that the chemicals used to replace CFCs, the main cause of the ozone hole, are creating additional environmental havoc too.

Some people also point to the redemptive power of philanthropy. Yet this too allows injustice, inequality, and violence to continue. Philanthropists pick the causes they like, overlooking countless others which may be more deserving, and frequently reinforce values and ways of life that they prefer. Powerful multinational companies also set up foundations or fund charitable programs which try to patch over social problems, easing the conscience of managers, shareholders, and consumers.

The existence of philanthropy allows bystanders to think "it's a shame about the poor/the burning rainforests/the preventable diseases in Bangladesh – but Mr. Rich and Megacorp are funding a school program for poor kids so something is being done." Charitable donors frequently impose conditions on their giving too – to discourage the use of contraceptives, or to promote their country's values or some religion, for example. While governments are subject to public scrutiny, the philanthropy business is not. Philanthropy and charity are substitutes for social responsibility because they postpone the reorganization of society into one where these problems do not exist.

Moreover, philanthropists rarely fight for social justice or reform. Their actions are typically founded on an acceptance of injustice, partly because they are its byproduct. It may be laudable for philanthropists to try to improve people's lives by offering medicines, encouraging research, financing schools, or bringing water to the poor. But most of this money promotes Western ideas of progress and development – and often explicitly so. It allows wealthy individuals and their philanthropic organizations to play God and decide what is worthy – without them being answerable to society for what they do. Genuine need should not be dependent on handouts at the whim of the rich.

By promoting economic individualism, the wealthy have also strengthened the case for minimal regulation. There is no sense in regulating companies, they argue, because businesses are slaves to the demands of the market, held hostage to the irrational behavior of consumers. It is consumers who demand the low gas prices which lead to oil spills that kill sea birds. It is consumers who demand the convenience of coffee cups that cannot be recycled. It is consumers who demand mobile phones with batteries that cannot be replaced so they need to buy a new phone every few years. This is all nonsense, of course. It is not people who demand these things. It is business owners demanding higher profits.

By encouraging people to think that they themselves always know what's best, and even to view scientific evidence as *just* another opinion, particularly in the English-speaking world, the modern belief system also neuters the voices of experts. By drowning out the views of the informed with those of an ill-informed majority, big business is able to act in its own interests without restraint.

As a result, the value of those who write in newspapers, post on social media, host YouTube channels, or front television shows is not determined by how insightful or clever they are but by how outrageous and controversial. The more clicks, traffic, and comment they generate, the greater the advertising they bring, and so the more their opinions are treasured. The result is that the obscure views of crazy outliers are often taken more seriously than those with something important to say. The equalization of opinions has allowed mass inanity to be commercialized, and very profitably too.

When it comes to climate change, this is a big problem, because the voices of those who know very little or nothing about the science of what is happening, as well as those who are paid to deny the truth, are regarded as equal to those who really understand the risks. As Richard Smith puts it, "by using democracy to dull consciousness, nurture complacency, and delay resistance, [economics has] succeeded in turning liberal democracy into zombie democracy, with devastating consequences."

Chapter 5

On the need to rethink everything

If democracies had been functioning as they should, climate change and environmental degradation would not have become existential problems. The warming of the planet, pollution of the oceans, and loss of so many species would not have been ignored. Citizens would have been properly informed about what is happening, and big businesses and the rich would not have been able to warp the political agenda for their own short-term financial benefit. The voice of the Green movement would have been heard and those elected to power would have acted in the long-term interests of all.

But modern democratic systems make the introduction of legislation to slow the pace of climate change almost impossible. The "tyranny of the masses" means that those who are informed about the risks cannot successfully push for change. Their voices are unheard over the cacophony of uninformed and false opinions.

The only way out of this dead end is through a radical overhaul of the entire social, democratic, and economic system. The existing system has to be torn down, so that a global network of governments of the people can be elected, which can be advised by a group of climate experts, and introduce the necessary regulations as fast as possible to avoid a global catastrophe. How they can do that is the subject of the next part of this book.

As well as tackling the problem of climate change, the challenge for governments is to change the way societies think about human development. They need to find a future pathway that does not demand the pursuit of endless economic growth and that does not encourage narcissistic individualism. This is not as difficult as it might appear, because there is nothing natural, eternal, or fixed in humanity's current world view.

The values that societies adopt do not happen by accident. They emerge and develop over time. They depend on the influence of the church, education, governments, and the media, for example, as well as collective personal experience. They can also be manipulated, as they have been throughout history, and as they have been in much of the world over the last 50 years. We will look at how this has been done in a moment.

This is another vital task for those wanting to build a better world. They need to open a global debate about the values a sustainable society should be built upon. It requires something like an Enlightenment or a Renaissance for societies to ask some basic questions that have not been thought about for a very long time.

It is *people* who decide what societies consider to be right and wrong. People select the indices of success and set the time horizons they consider useful. People define how they relate to one another and how societies relate to nature. What people consider to be their entitlement, or their duty, what they think of as freedom, and order, are all choices. If humanity is to release itself from its self-induced ecological burden it first needs to understand that the dominant system of human development used in most of the world today, also known as neoliberalism, is the main cause of their problems.

Part 2

Got it! What do I do?

Chapter 6

Prepare for a fight

Imagine an army driven into retreat, rapidly pursued by its enemies. There is a growing sense of fear in its ranks. After some time, the army comes to a wide, fast moving river. What lies on the other side is unknown. There is no other path, and no bridges or boats to get to the other side. Doing nothing will bring disaster. The army has to find a way to cross the river, even though it seems impossible and the risks are very high. There is no choice.

That is where societies find themselves with climate change. Societies have to find a way to leave where they are and move to a strange and unfamiliar place. Getting there will not be easy, but doing nothing brings certain disaster.

At some point enough people will realize this. We hope that realization will come soon, as a result of Covid-19. But it may not. It may come later. But, just like that army, enough people will eventually see that they cannot stay where they are, or continue to live as now. They will realize that humanity has reached an impasse and must find a different way. That is when radical ideas for an alternative approach will be welcome.

It is our most fervent hope that this day comes soon.

When societies are ready to consider a different model of human development, what will they need? That is the subject of the second part of this book.

Societies will need a blueprint for a different future. To slow the pace of climate change, the new system must be based on humanity living in balance with nature. People will need to live in better balance with each other too. To be sustainable, this new, clean-sheet future must be based on principles of equality – financial, sexual, racial, and equality of opportunity – but also with the recognition that not every *opinion* should be treated equally when it comes to making wise, informed decisions in the interests of the majority.

While many of these conditions might initially seem like nice-to-haves, they are not. Creating a more balanced approach to human development is the only way the transition will work. Dignity and equality must be hard-wired into the new system.

Societies can't just tweak the system

Most countries have been riven by inequality and injustice for decades. The transition that lies ahead requires them to fix this. If societies do not eradicate these injustices, the future system will be unstable, which will make it impossible to successfully slow the pace of climate change.

People have worked tirelessly to grind down the harsh edges of the current system for years, to make it fairer, to make it work more effectively, to protect the weak, and prevent its many excesses. Most of those efforts have been well meaning, though not all. Some were cynical and designed to deceive. Yet all those efforts to make the system kinder and less damaging were mostly wasted because those working for reform were effectively trying to turn a wheelbarrow into a spaceship.

As we will explain in the next few pages, the dominant system of human development used in most of the world was specifically designed to maximize short-term profits for a small, wealthy group of people. An externality of the system is that it creates climate change, just as eating too much sugar rots your teeth. It is impossible to stop climate change within a system which focuses on increasing short-term financial gains for a wealthy few, because it is the push for growth and profit that is the source of the climate change problem. Likewise, the system cannot be made fairer for the majority, because it is the major source of injustice.

As human rights activist Malcolm X put it during his fight

against injustice and racism:

> *[I]t's impossible for a chicken to produce a duck egg – a chicken just doesn't have within its system [what is needed] to produce a duck egg. It can't do it. It can only produce according to what that particular system was constructed to produce.*[15]
> **Malcolm X**

The current system serves the interests of a small group. It excludes the majority of people, crushes new ideas, and opposes any other purpose. It cannot produce equality or ecological sustainability for humanity and all the other species living on earth today. It can only produce what it was designed to produce. And that is ever-higher short-term profits for the wealthy. This means that those who gain from the system, and so have power, do not want change, though they might sometimes say they do. Those who benefit from the system are incapable of reform. They do not know how to reform. They only know how to extract, to exploit, and to reproduce the conditions of inequality, ecological ruin, and injustice.

For change to come, the societies of the future need to be led by completely different people from those who lead them today. They need to be led by people who are mostly in their 20s or 30s, who have the most to lose, as well as by people who have not been brainwashed by the current system, and can imagine a different, less selfish approach to human development.

For decades, the vast majority of people have obediently gone about their lives, doing what the system expected them to do, doing their jobs, looking desperately for jobs when unemployment came knocking, voting Red or Blue or Green when they were told to, and believing that some intelligent, benevolent authority had their back. It never did.

Under the current system, money has always flowed *away* from the majority, while the laws which they thought might protect them have been used to persecute them instead. Meanwhile, the sky has turned black with burning oil fields, the oceans have become slick and shiny with oil spills, and hurricanes and wildfires have destroyed people's homes.

In the harsh reality of the *non*-inclusive neoliberal capitalist world, Covid-19 has mostly killed the poor, and those who the system has made chronically sick through overwork, stress, and low-grade food. The virus has killed the street sweepers, hospital cleaners, meat packers, immigrants, and minorities in vastly greater numbers than it has killed political leaders, the rich, and the world's CEOs. That is not down to the victims' bad luck.

Modern legal systems, which often started with a desire to treat everyone equally, have become so corrupted that they reinforce injustice and inequality by protecting the powerful while punishing the weak for even tiny transgressions. George Floyd, a black American from an economically deprived background, was arrested in 2020 for allegedly using a fake US$20 bill. He died fewer than 20 minutes later, choked to death by a white policeman kneeling on his neck, handcuffed, face-down, pleading for his life. Within days, Roger Stone, an American conservative political consultant and lobbyist, who had been convicted of obstruction of justice, tax fraud, blackmail, bank fraud, lying to the US Congress, and witness tampering was set free by his good friend, President Donald J. Trump.

Most people are required to obey laws and, if they are convicted of a serious crime, serve time – if the system doesn't kill them first – while the wealthy and well connected are too often set free, even when a court finds them guilty. The bankers who fueled the financial crisis of 2008 or stole from Malaysia's sovereign wealth fund have not faced any serious penalty, just as the super-rich and politically powerful have been able to set up shell companies in tax havens to avoid paying what they owe to the societies that disproportionately sustain, support, and protect them, the very societies which provided their massive, unreasonable wealth in the first place. It is the same thinking which has allowed companies like Shell and Exxon to lie about climate change for 40 years, while generating vast profits for their executives and shareholders from the destruction of ecosystems and the planet.

Chapter 7

System change not climate change

Without going deeply into the theory of systems change, if there is to be a relatively painless transition from the current system to a better one, two conditions need to be met.

First, the existing system needs to have lost widespread legitimacy and fallen into a crisis. Its failures must be obvious to a large number of people. (Condition 1: check.) With growing environmental problems, widening inequality, an unjust legal framework, and unsustainable levels of debt, the current system had lost much of its legitimacy even before the onset of Covid-19. Since the virus began its spread, the cracks in the system have widened. Many governments failed to protect their citizens' physical and financial health with some revealing themselves to be incompetent, uncaring, and violent, only interested in protecting wealthy interests. In the UK, Brazil, and US the needs of businesses were put before the lives of people.

Second, there needs to be an alternative system waiting in the wings, ready to take center stage. (Condition 2: pending.)

This is what those who want radical change need to focus on: the development of a framework for a different system which can be ready to deploy when the time is ripe. The framework for the new system does not have to be perfectly designed in every detail. It just needs to be sufficiently well conceived to be credible.

This approach to systems change has plenty of precedent, from the CIA-managed transition of President Sukarno to President Suharto in Indonesia in 1967, to some eastern European countries moving from Soviet administration to the European Union during the early 1990s, and China's *de facto* takeover of Hong Kong in 2020. The approach also has precedent in America and Europe.

Following the Second World War, as Keynesian economic ideas became discredited, a group of neoliberal economists started work on the alternative system. In the late 1940s they formed the Mont Pelerin Society and with generous funding from corporations and wealthy individuals they developed an alternative approach to social and economic development. Over the following years, they spread their ideas and increased their influence by building networks at universities, through the creation of dedicated think tanks, and by publishing academic papers, news articles, and books.

Their chance for fame came in the late 1970s and early 1980s, when Margaret Thatcher was elected Prime Minister in the UK and Ronald Reagan became President in America. When these politicians looked for alternative ideas for managing societies, to make a break with the past, they were presented with the work of the Mont Pelerin economists, including Friedrich Hayek, Karl Popper, Ludwig von Mises, and Milton Friedman – some of the twentieth century's best-known economists.

Most of the leading economists in the Mont Pelerin Society are closely aligned to what is known as "the Austrian School of economics" because their families have roots in the Austro-Hungarian Empire. Propaganda specialist Edward Louis Bernays, whose family also came from Austria, and who is better known as "the father of spin," also played an important role in spreading Mont Pelerin thinking. He worked on redefining modern ideas of democracy to promote individualism, and undermine common ideas of social cohesion.[16] Both concepts are central to Austrian School thinking. Originally, the term "Austrian School" was intended as an insult, because mainstream economists saw the ideas of this group as outcast, obscure and provincial. Despite this, they succeeded in having their approach widely adopted around the world.

It is the ideas of these economists which have dominated

social development thinking in most countries for nearly 40 years and which have led humanity into the dead end it finds itself today.

The Society's goal was to spread neoliberal thinking and what the group claimed were the "central values" of civilization. They wanted people to believe that government and the welfare state were dangerous, that each individual was sovereign, and that businesses should be minimally regulated. President Reagan's quip that the nine most dangerous words in the English language are "I'm from the government, and I'm here to help" was deliberately designed to undermine the role of the state and spread the apparent logic of these ideas.

Of Reagan's 76 economic advisers, 22 were members of the Mont Pelerin Society. Thatcher's chief economic adviser, as well as many other economists close to her, were members of the Society too.

For more than half a century, the Mont Pelerin Society has worked hard to make people's belief in the free market so normal that, like a religion, the underlying ideas are rarely questioned. As well as working to change the way economics is taught in universities and schools, the group has carefully built links to senior politicians, bankers, and journalists.

It was also behind the creation of the "the Sveriges Riksbank Prize in Economic Sciences in Memory of Alfred Nobel," more commonly known as the Nobel Prize in Economics, and which started in 1968. This is not a real Nobel Prize, like those for literature or science which date back to 1901. It is a public relations coup for free market economists, designed to legitimize neoliberal thinking.

The Society has also cleverly sought out intelligent people who are not economists, as secondary carriers of their ideas. With targeted articles, documentaries, and news reports to promote neoliberal free market thinking, many well-respected members of society have become inadvertent advocates of the

Mont Pelerin's work and enthusiastic supporters of the neoliberal system. Because they trust the lofty academic credentials of the Society's economists, and because they have not suffered from the system's many flaws personally, these people do not question the underlying assumptions or examine the outcome for the majority. The Society is also greatly responsible for "mathematizing" economics, to make the subject appear rigorously scientific and so give it greater intellectual credibility.

Despite the damage it has wrought, the Society provides a good model to follow. The work of this small group of economists shows how a new system can be defined, legitimized, and communicated. Unfortunately, the Mont Pelerin approach also carries two big warnings. First, it shows that it is possible for a small group of people to successfully promote a system of human development which is based entirely on self-serving, half-true ideas. Second, it shows that a system of human development does not have to deliver what it promises. In her excellent book about the modern history of neoliberal thinking and the role of the Mont Pelerin Society in creating the current economic system, Chien-yi Lu writes:

> *Immeasurable amounts of time, energy, and talent have already been wasted on engaging [in] sincere debate with neoliberals as if they were honest theorists, thinkers, scholars, think-tank experts, or statesmen when actually, the core feature of neoliberalism is deceit.*[17]
>
> Chien-yi Lu

If humanity is to slow the pace of climate change, the new system of development needs to be based on foundations that are robust, unlike those of neoliberal economics. It must also have clear operating parameters that are consistent with the achievement of the primary goal – to stop runaway global warming.

Chapter 8

Blueprint for the future

So, what does an alternative model of human development actually look like? Compared to the current system, it is almost unrecognizable.

Below is a rough sketch of the system needed, though it would not be possible to introduce it immediately. It is a blueprint to work towards, not a plan to introduce next week. To reduce the impact of climate change and create a world where humanity can live in balance with nature, societies need a system that is free from greed and materialism, with genuine equality. Developing that will take some time. Note that we have not considered how democratic a sustainable system should be.

1. Respecting nature

A sustainable system must first respect nature. Resource use and pollution need to be strictly limited, with a fixed maximum human ecological footprint regardless of the population.

What people currently consider to be sustainable – wind farms, solar panels, and electric cars – are not sustainable at all. They use vast quantities of scarce raw materials and energy, and have lifespans that can be measured in decades, at best.

Sustainable societies need to "fit," as Herbert Spencer observed. When he talked about the "survival of the fittest" (the term is often wrongly attributed to Charles Darwin), he did not mean that competition is good and that only the strong survive, as free market economists claim. He meant that the species that prospers is that which best "fits" its surroundings. Survivors are those that are best adapted to live in balance with the world around them.

2. Meeting everyone's basic needs fairly

A sustainable system must meet everyone's needs for food, safety, purpose, mobility, communications, and shelter, and it must do this fairly. This is necessary to sustain life and to eradicate injustice. It minimizes conflict, promoting resilience and stability.

Think about it this way: in every community there would be a place where everyone could choose from a wide range of vegan dishes three times each day. Like the housing, this would be available for free. Agriculture would be localized and only use natural fertilizers. Renewable electricity would be the main source of energy (though not generated as today). Almost 100 percent of the resources used would be in accordance with the principles of a circular economy – recycled, renewed, repaired, and reused. The use of scarce non-renewable resources would be infinitesimally small.

Once society has met the basic needs of all citizens, rewards for individual achievement would be possible, so long as the gap between those who are treated differently and everyone else is very small, and as long as any achievements and individual needs are recognized justly. As well as a guaranteed minimum living standard, there would need to be a maximum living standard too.

3. Equal rights for future generations and other species

The system must guarantee the rights of all other living creatures to prosper, and regard the rights of future generations as equal to those alive. No generation can be allowed to damage the natural environment if this disadvantages future generations or other species. Each generation must leave the planet in the same condition they find it, or better.

4. Leisure time to offset productivity gains

A sustainable society would still need to innovate. It would need to reduce waste, improve the rate of recycling, increase energy efficiency, and pursue advances in medical science, for example. Services would need to innovate too, to better manage the process of sharing what is produced.

To limit the human ecological footprint, efficiency gains would sometimes need to be offset with increased leisure time. If a factory develops a new technology and can produce more, it may have to give its workers more vacation time rather than increase output, to avoid excess production or waste. The incentive for people to innovate would be the knowledge that their work improved human well-being.

5. Without money so everyone is equal

As with every other species, a sustainable system would function without money. It is a myth that money was invented to make trade easier. It was created to fight wars, so that armies could be paid.[18] It has been maintained since as a form of oppression. Wars were fought to access the metals needed to make coins. It justified the plunder of South America. The creation of debt has led to misery for millions.

Eradicating money removes one of the biggest sources of resentment, conflict, and injustice. A sustainable system must be based instead on temporary rights of stewardship to commonly held land or property with long-term social accountability.

Getting rid of money has many advantages. It extinguishes banking, speculation, and property ownership as well as all forms of accumulated wealth. It removes the possibility of people inheriting an unfair start in life. And it puts an end to inflation, debt, and taxes. All are extremely divisive human constructs, which have allowed a minority to oppress the majority through conflict, enslavement, and debt for generations.

6. Improved human well-being is the goal

Rather than boosting output and consumption, humanity

would focus on developing artistically, culturally, and intellectually. It would boost well-being and life satisfaction, as well as physical health. Sports and religion can flourish, as can scientific research. It is only the raw materials flow that must be kept below a maximum sustainable level, so that scarce non-renewable resources are not depleted to any measurable degree and environmental degradation never breaches natural limits.

7. Population constrained

The human population could increase in a sustainable system, restricted by the availability of scarce resources and the level of pollution. Every additional person would lower the average living standard for all. Put simply: a small population would live better than a large one.

8 Common Good responsibility

The new system would prioritize the long-term common good. Individuals and businesses would not be able to breach nature's limits by using resources unsustainably, generate more pollution than nature can absorb, infringe the rights of future generations and other species, or create any social injustice.

Of course, many people will find these ideas difficult to understand, certainly at first, and this will slow the rate at which they can be introduced. So, change-makers will always need to maintain a focus on the first-priority objective, which is to slow the pace of climate change as fast as possible.

In the long-term, a system based on these ideas would lead to happier, more fulfilled, and more stable societies. It would create a future for humanity based on dignity and harmony, in balance with the world.

When asked to paint a picture of what this sustainable society might actually look like, our minds wander to life on Jean-Luc Picard's starship Enterprise, and also to some ancient civilizations.

Life in a sustainable society is based on mutual respect, with everyone fulfilling their assigned role. Those "in charge" are really just conveners who bring others together, to collect individual ideas and insights. Informed decisions are made collectively, in the interests of all. There is no need for money because everyone is equal. Respect for nature and each other comes first. Life satisfaction comes from a job well done for its own sake, not the praise of others. It comes from education, artistic endeavor, deeper understanding, and the thrill of exploration.

Life in a sustainable system is a secure voyage of discovery, not an endless treadmill to pay the next interest installment.

Chapter 9

Preparing for power

We said that change-makers need to develop a blueprint for an alternative system of human development and provided a sketch of what that could look like. To develop this framework further will require experts in development, as well as those with an understanding of human behavior, and specialists in climate change. It will require a group of wise people to come together and think the system through. They will need to examine the implications for different peoples and cultures, and work out how to shutter the fossil fuel industry, the auto business, aviation, and all the other polluting businesses in an orderly way. As with the Mont Pelerin Society, this group might perhaps consist of a relatively small number of like-minded people, certainly at first. Unlike the Society, however, those who participate in this group must be willing to put the long-term future of humanity before their own self-interests. To us, it is very unlikely that any of the people in this group will come from anywhere near the top of the current system. We will discuss this more later.

One way to set this process in motion is through the creation of a global political party for change, which can directly challenge the existing system. This new party must have just one principle goal – to slow the pace of climate change as fast as possible by cutting greenhouse gas emissions. It might build on the momentum of organizations like Extinction Rebellion, many well-established Green Parties and Fridays for Future, to bring these battle-hardened protesters into an organized political movement for change.

This group should first try to use the existing democratic process to replace the current development system with a new and sustainable model of human development. As well as pushing for radical systems change, to maintain its support and legitimacy, the new party should push governments to give financial and emotional support to those who will need it during the transition. It should push for the prosecution of those who have been responsible for the climate crisis too. This is not simply in the interests of justice. It will help everyone understand the rottenness of the current system and see more clearly the damage that has been wrought.

The group will need to:

1. Form a political party with one main goal – to slow the pace of climate change by rapidly cutting greenhouse gas emissions.
2. Pursue systems change through the current democratic process.
3. Push for the rapid closure of polluting industries everywhere.
4. Push for agricultural reform and a halt to deforestation.
5. Work to provide financial and emotional support to those who will need it during the transition by pushing for the introduction of a basic lifetime income for all adults (in a way that does not encourage further population growth).
6. Prosecute those who are responsible for the climate crisis.

Societies need to understand that without radical change, billions of people and other species will suffer terribly, and there will be no way to turn back. A key element of the battle will be to deepen understanding about the climate problem. Those seeking radical reform need to follow the example of the Mont Pelerin Society, and influence those in education, in politics, and the media.

Instead of being offered an alternative development system, those in power need to be told instead that they have failed and need to go. Today's political leaders must make way for others.

There cannot be any compromise on this. If they were to gain a foothold in any coalition for change, they would work tirelessly to kill it. Those who currently lead our societies need to step aside in the long-term interests of everyone. They are not what humanity needs, and never were. More on this later.

We know that what we are proposing will be very hard to pull off. It might seem idealistic and improbable. But we have also worked on this problem for long enough to know that there is no other way ahead. Covid-19 has shown societies that radical change on this scale is not as impossible as many once thought. Humanity is like that army by the river. It cannot go back. It has to find a way to the other side, and then live differently. It has to push forward and make a transition, no matter what this means economically, politically, or socially. And it has to keep pushing ahead no matter how improbable reaching the goal might sometimes seem. Societies must find a way to make the crossing to another system or collapse in the attempt, because doing nothing will destroy everything.

Thankfully, there are plenty of examples to show the way ahead. There are countless examples of societies from the present and the past that show that humanity can live in much better balance with nature. There are others that prove that people can work together for the good of all. Covid-19 is the opportunity that makes a peaceful transition possible.

There *will* be a transition to another system. It could be to the one we are proposing here or it could be more confrontational and chaotic. At some point soon, however, the deep inconsistencies in the current system and the effects of climate change will come together and force change.

For now, it is still possible for people to choose.

Chapter 10

Obstacles on the path ahead

There are many obstacles that block the change that is needed and they will not be obliterated by having a good plan. Life is not like that. But those pushing for change can prepare for the resistance. To do this, new thinkers will need to shine some very bright lights into dark, incompetent corners and reveal the truth.

Revealing the truth is important, as is the ability to move fast, adapt to circumstances, and change course, though always with an eye on the end goal. Those on the other side, who want to prevent change, will be versatile and resilient too. They also have money, power, and authority to back them up.

Who are these people who block the change that is needed? Sometimes they are easy to spot: the suited oil exec getting into his helicopter is an obvious icon. But sometimes the people who don't want change look and sound like everybody else. Some look like powerful politicians, whizzing about in black limos, with harried assistants briefing them on their next press conference. Some were born to money and power. Some look like your sister. Some just worked damned hard, got through grad school, and finally landed a great job in a hedge fund six months ago.

Some of those who block change actually *are* responsible for climate change. They decided to build all those coal-fired power stations, manufacture products that cannot be recycled, and fit container ships with extremely polluting engines. They invest in the oil business, knowing it is environmentally destructive. They build airport terminals knowing that they will cause more air pollution. They build car factories knowing that the energy required and the pollution created will damage the atmosphere. They work in product design, marketing, and sales to persuade everyone else to buy destructive products. They work for the public relations firms and think tanks that spread neoliberal ideas and doubts about climate science.

Environmental destruction and climate change are not happening on their own. Most of the people who make these decisions know, deep down, that they are doing something wrong. They know that they are causing climate change. They know that more flights and car journeys cause more pollution. They know their plastic packaging will poison the seas.

You didn't create this problem

Winning the climate change battle is not about the rest of us turning vegan. It is about the majority of people on the planet stopping the destructive behavior of the small number who are the problem, and taking back democratic control so that they do not destroy everyone's future. This is not a battle between the 99% and the 1%. It is a fight between the 99.99% and the 0.01%.

There *are* people who are directly responsible for climate change and it is the activities of these people that must change. The task for the rest of us is to make that happen.

In many ways, the behavior of these people is curious.

At one stage in the story of Leo Tolstoy's book *Resurrection,* a large group of prisoners is forced to walk to a railway station where they are to be transported to Siberia. It is baking hot, and the prisoners are outside for hours without water or shade. Several die from heatstroke.

Tolstoy argues that these prisoners did not just die: they were killed by the decisions of prison wardens, guards, and the judicial system which exiled them. Their deaths were not an inevitable result of any crime they might have committed. The system killed them.

They were killed because of the decisions of other people, and yet no one was held accountable. The reason, said Tolstoy, is that those in charge were able to believe that "there are situations in this world in which loving consideration for humanity is not obligatory." The people responsible for these prisoners' deaths would have behaved differently if it had been their families they were sending to the station. They would not have sent out their relatives in such heat. They would have provided shade and water, or waited until it was cooler.

The guards abandoned their compassion because they viewed the prisoners as less than human, and because they valued the expectations of their jobs more than their humanity.

The same has happened to those who work for the companies that are polluting the planet. Employees of the fossil fuel industry, the cement business, car manufacturing, and so many other companies which are damaging the planet, have lost their sense of humanity, and their love for those around them. They have become numb: they have forgotten their duty to everyone. Those who pilot planes and build car factories would never put the lives of their families in danger, or risk the future of their grandchildren. Yet they do it collectively every day, hidden behind their sense of corporate duty which makes what they do seem okay when it is not.

Sing the la-la song

There are plenty of others who don't want change, mostly because, to some extent or another, they benefit from the system as it is. For them, there's no reason to change – even if they are told that they will become pot roast like everyone else. They block out apocalyptic messages, even when there is plenty of evidence that an economic and ecological tsunami is on its way. Somehow, it's just too hard for them to imagine a future that isn't a simple extrapolation of how they live now. Covid-19 illustrates this. Even when infections were spreading quickly and the death toll was mushrooming, millions of people were pushing for life as it was before. They could not understand a non-linear change. Humans have a fantastic ability to sing the la-la song.

Cartoon – Singing the la-la song

For this reason, we invented the la-la index. The la-la index is a measure of different countries' or regions' determination to stick their fingers in their ears (metaphorically of course) and sing the la-la song to drown out reality. It can be applied to any difficult reality, from a child being told to go to bed, to people's response to Covid-19, or global society facing the threat of catastrophic climate change.

In the example below we have applied it in a few countries to compare how they responded to the threat of the Covid-19 virus pandemic in the first half of 2020.

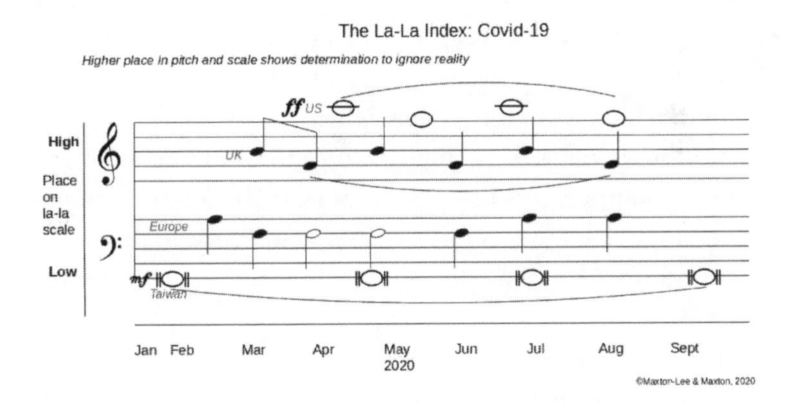

La-La Index – Image

Taiwan is rock bottom on the index: it could not sing the "la-la" song at all. As the virus developed, Taiwan took rapid and extensive measures, shut the country to travelers, and compelled all citizens to wear masks in public, wash their hands, and work from home. The Taiwanese started early, applied the measures consistently, and were ultra-cautious about lifting them. At the time of writing, eight months on, it was still not possible to leave and re-enter Taiwan easily.

Europe, the next region on the index, was slower to respond to the threat, continuing to allow inter-regional travel even while Italy's cases rocketed. Then lockdowns were suddenly applied with very little notice, once authorities recognized the speed of the spread. Measures were reasonably strict, although a culture of selfishness weakened the effects in some places. It was only really in the summer months of 2020 that Europeans began to sing the la-la song in earnest, when the desire for going on vacation overrode the fact that people were still getting sick. The obvious exception to the general European trend was Sweden, where the la-la song was more popular than Abba. The Swedes did almost nothing in response to the spread of the disease and had among the highest death rates in the world, per head of population.[19]

The UK and US register very high on the la-la index, reflecting an astonishing capacity to ignore a visible, looming threat, even as the data and experience in other countries mounted. The US and UK showed an impressive ability to pretend that the suffering and death of their own people was not really happening. They were able to stick their fingers deep in their ears and sing the la-la song very loudly, flip-flopping on policy, changing their minds about strategy and advice, and generally weaving about like a drunkard at the wheel of a pickup truck.

The la-la index is not absolute, of course. A low score in one situation does not necessarily guarantee a low score in another. But it offers some useful insight into how well societies are able to accept and respond to factual information. It was just too hard for the high la-la index countries to imagine doing something as radical as shutting down their economies. Hundreds of thousands of businesses as well as millions of people whose incomes, rent-paying, loan-repaying, and life-planning depended on keeping the economy open fought against the scientific evidence, even though it caused thousands of deaths.

The failure of people to consider alternative solutions in times of trouble is another obstacle to dealing with climate change. If the key message clashes with popular thinking, many people sing the la-la song and refuse to pay attention. Those working for change need to understand this and deliver a strong, consistent, easy-to-understand counter narrative which can educate people quickly and help them think differently.

Change-makers need to show that alternative solutions exist. At the start of the Covid-19 crisis, the governments of many countries paid those who were furloughed so that they could get on with their lives. This was difficult, because it required a change in thinking. Governments had to pay people to stay home and people had to take payments without doing any work. Some governments fought against the program, fearing the economic consequences, and that people would become dependent on handouts and not want to work.

Yet this is exactly what will be needed during the transition. Those who lose their jobs during the shift to a sustainable system need to be paid generously by the state, so they welcome change. Governments will also need to accept that some of these payments will need to be made for years. People will need to be retrained to work in a post-fossil fuel world. It will take time for this to happen and for new business sectors to emerge.

So where is the money to come from?

Pay as you go

That's easy: Governments can print money. After the 2008 financial crisis, governments around the world printed trillions of dollars, pounds, euros, and yen to save the banking system. If they could do it then, they can surely do it to prevent societal collapse now.

Printing money on this scale could cause economic problems, of course, though not much happened after governments cranked up the printing presses to save the banking system after 2008. There was no surge in inflation or collapse of currencies. Debt increased, but debt, like money, is not real. It is just numbers on computers which also have delete buttons. Ignoring debt can lead to conflict, so it has to be diplomatically managed. But even if printing money creates economic problems and a buildup of debt, it will not inevitably lead to the deaths of hundreds of millions of people. Uncontrolled climate change will.

Emotional fallout

As well as supporting people financially during the transition, there will also be psychological fallout from shutting down unsustainable industries. Many people who were furloughed because of Covid-19 suffered mental health problems, and a loss of purpose and identity. There are two major issues here: the trauma of job loss, and an understanding about why the cuts are necessary.

For most people, losing a job is traumatic. It can feel like a personal rejection ("I'm not needed") which is deeply wounding. It can be like a bereavement, every bit as painful as a relationship breakup. For those who have worked in the same place for a long time, their colleagues are often friends. They may even feel like family. That is hard to leave behind. The loss of seeing those people every day, in the same setting, can be painful and disorienting. Most people have strong feelings of identity around their jobs, which are uprooted when they leave, and this can be bewildering.

People who have lost their jobs (and those who have retired) often wonder "Who am I now? *What* am I?" If not carefully managed those questions can run into a deeper, more existential crisis with the person questioning the point of life itself. Many people turn to drugs, alcohol, or gambling.

Strongly related to the question of identity is the bubble of "I'm doing okay" that everybody, to some extent, wraps around themselves. It's a natural, human, trait for dealing with life's wild range of difficult situations, from "I really don't want to spend Christmas with Uncle Simon again" to "I know deep down that this oil company I'm working for is killing people."

Of course, nobody wakes up and says "Yay! Time to wreck the planet and murder innocents again!" Not even people working in the so-called "defense" industry or the armed forces. People tell themselves there's meaning and purpose in what they do. When it comes to the climate transition, this is a real problem because it makes change harder. Take Norway as an example: 4% of Norway's population works directly or indirectly in the petroleum industry. Imagine the personal psychological trauma for each of them, those brothers, mothers, fathers, and aunts, if they really had to accept that their work is killing people. To face a truth like that is a serious thing. It would shake the foundations of a person's psyche. Large numbers of well-trained care workers will be needed to help manage this emotional journey.

These psychological effects can be as real as a physical wound; a broken heart hurts as profoundly as a broken leg. The new system will need to have strong arms to catch those in pain as they fall, to help them understand why this is necessary. It will need to help them heal and move on, and support them as they find a new identity and purpose.

There is, however, another level of resistance quite different from the inability to hear uncomfortable realities or make difficult transitions, and it is a formidable, frightening, and sinister resistance. There are the companies and groups of people we mentioned before – those who are actually responsible for the catastrophe everyone faces. These people know very well what is happening, and yet they deliberately twist the truth, because they want to profit from the current situation. Rather than accepting the need for change, their goal is to reinforce the system that has brought them so much wealth and power.

Major oil companies (including Exxon, Mobil, Amoco, Phillips, Texaco, Shell, Sunoco, Sohio, and Standard Oil of California and Gulf Oil (two companies that became Chevron)[20]) knew in the 1980s that burning fossil fuels caused climate change, and they knew how dangerous that situation would become. Today, these oil companies run expensive branding campaigns saying they support action against climate change, while at the same time greatly expanding their fossil fuel extraction businesses. They also spend hundreds of millions of dollars to block, delay, or undermine policies that would actually tackle climate change.[21]

The people that control and finance these businesses are some of the wealthiest in the world and they know how to control information. One tactic they use is "astroturf" campaigns, so called because they look green but are fake. These include the use of fake consumer or grassroots movements that are funded and directed by the fossil fuel industry. The tobacco industry used the same trick many years ago. They deliberately deny and discredit facts, while giving the impression that certain destructive products or actions hold extensive public support. There are many other clever and divisive tactics they use and they will continue to present a huge obstacle to change. The leaders of the new system need to prepare for this very well-funded, powerful resistance. They will need to identify and fight the lies.

The Judean People's Front obstacles

Another obstacle is the risk of infighting among those pushing for change. It is tempting to believe that, if people unite around a good cause, they will work together to achieve that goal, and not be distracted by petty differences or jealousies. It is nice to think that the objective of a better life for everyone, which prevents suffering and destruction, will be enough to make everyone pull in the same direction. In reality, it is rather more complicated.

Too often when people come together to work towards a common goal, they become like "the People's Front of Judea" in the 1979 Monty Python movie *Life of Brian*. In part of the story, a small group of passionate anarchists are regularly seen plotting to overthrow "imperialist" Roman rule. They talk about revolutionary change and get very stirred up. But then they get bogged down in their own internal disagreements and procedures. They talk endlessly but never actually do anything.

Coming together to fight against something as important as climate change, people might think their differences don't matter, but conflicting visions of what is important can create big divisions. It is important, therefore, for everyone to keep the big picture front of mind. There is only one goal: to prevent runaway climate change. There is only one way to do that: to reduce greenhouse gas emissions by 60% within 10 years, and then to zero. Other social problems might be made better by doing this, and by the establishment of a new system of governance. But they must not take priority. The one central goal must remain clear at all times.

Related to this is a human tendency to attack each other instead of being united. The People's Front of Judea absolutely hated other, almost identical, resistance groups such as the Judean People's Front, and the Judean Popular People's Front, even though they were fighting for exactly the same cause. This twist in the movie is important to remember. Societies need to remember that the common enemy is carbon emissions, and the human activities producing them.

This means that people will need to find new ways of working together. In the beginning, it will often feel exciting, and maybe a bit romantic. But as communities and societies get down to the not-so-romantic jobs of building a new system and making it work, the shine will start to rub off. Ideologies will start to conflict. Personalities will clash. People will need to dig deep and find their solidarity, to see beyond irritations, dislikes, and envy, and stay focused on the goal.

Much can be learned from cooperative societies like those in Asia and in parts of Europe. In this case, we are thinking about places like Taiwan, Germany, France, Austria, and much of Scandinavia. Some of these countries have had their share of trouble and trauma but, perhaps through these problems, their citizens have built a culture of interaction that is mutually respectful and focused on a common goal. Discussions are not about one side winning at the expense of another. They are solutions oriented, not confrontational, as so often in the English-speaking world. Of course, conflicts do occur, but when the common approach is about finding the best solution for everyone, people usually find a way of working together.

When it comes to identifying the best solution, old truths and priorities will die hard. Things which seem like common sense (don't stick a metal fork into a toaster; don't lick a frozen lamppost) are generally not truths that humans are born with: they are ideas that are taught. Their world view is taught too. Apart from the real essentials of everyday life (food, clothes, love, somewhere to live, and sleep), the priorities people have from one day to the next are mostly formed by society. This makes it hard for them to understand that the way they see the world has been manufactured and is not fixed. This makes it harder for them to understand when priorities need to change. But that is what has to happen.

Largely because of the success of the Mont Pelerin Society, people believe that economic growth, individual responsibility, and minimum regulation are the only way. It's just like a feudal, Medieval society where a serf blindly follows the system defined for him, unable to imagine any alternative.

He gets up with the sun, works for his lord, gives away most of what he reaps, and starves when he can't pay his tithes. It's the natural order.

The belief in today's system goes very deep. People think it is normal for companies to increase profits every quarter. They worry when economic growth slows or stock markets decline. This makes it hard for them to think outside the box of "how we do things every day," and yet societies are going to have to re-learn almost everything. The pursuit of endless economic growth and short-term profit maximization is deadly. It is the cause of climate change and widens inequality. Yet most people keep chasing the pot of gold at the end of the neoliberal rainbow, because trillions of messages have been injected into societies like a virus, through advertising, education, TV, social media, politicians, and corporate branding. Societies are infected with the drug of "more money, now!" and they need to go cold turkey.

New thinking needs to shine a painfully bright spotlight on these neoliberal lies while the people who are responsible for climate change need to be prosecuted – the heads of the fossil fuel companies, airlines, and automotive firms as well as their shareholders, and all those who have knowingly created these damaging gases to boost short-term profits. The people who financed these businesses or supported them through marketing and advertising, as well as all those who sought to deny what was happening, or make it seem okay should be prosecuted for crimes against humanity and nature too.

Of course, these companies have large, sophisticated legal teams who have made it difficult to hold the people behind these terrible actions accountable. Those responsible have been able to hide behind the inanimate, faceless shield of their company. So, the legal firms which have helped to protect these people, and the law schools which taught students how to weave lies and deceit into the fabric of global society, should be named and shamed and prosecuted for their crimes too. The Public Relations agencies and merchants of doubt should also be held accountable. The actions of these greedy, conscienceless people, have brought catastrophe to all. Why should they be allowed to walk free, as if what they have done is okay?

The selfish, damaging actions of these people, and many others, are frequently defended in Western societies under the banner of "freedom." But there is a difference between the individual freedom which causes no harm to others and what these people have done. Right now, societies are still defending the right of individuals to harm others for their own short-term gain, because the logic of individual freedom compared to the greater good, or harm, has become confused.

We, not I, nor me

Because of the way societies have been taught to think, many people find it very hard to imagine giving up their individual freedoms. People guard them, sometimes irrationally, ignoring the harm their actions inflict on others. A clear example of societies putting individual freedom before the well-being of others was with the thorny subject of mask wearing during the early stages of the Covid-19 pandemic. This became so extreme that there were demonstrations against mask wearing on the streets of Madrid, while, in France, a bus driver was murdered for telling his passenger to wear one.

Pre-Covid, people in Europe and the US would feel offended if they saw Asian tourists wearing masks. They would mistakenly imagine these tourists were worried about air quality or fret about getting sick. "Our air is clean!" protested a friend of ours in Vienna. "Our hospital is hygienic!" said an indignant doctor at the beginning of the pandemic in Minnesota, when a Taiwanese family came to visit their daughter's newborn baby.

Even as thousands were dying every day, many in Europe and the US objected to wearing a mask while, in Asian societies, there was no problem. Even before Covid-19 people were used to wearing masks in Asia, when they had a cold or a cough, or if their sniffles were due to allergies and not contagious. Why? Because in Asian societies people want to show respect and concern for people around them.

Wearing a face mask in Hong Kong, South Korea, Thailand, or Taiwan, says "it is important to me that you feel comfortable and safe. Your well-being is as important as my own." To an extent, it actually says "your well-being is MORE important than my own" because people are willing to make themselves a little less comfortable (by wearing a mask on hot days, for example) in order to secure the well-being of others.

Refusing to wear a mask during the pandemic said "my freedom is more important than your life." It said "I am more important than you, your elderly relatives, or your community." It said "I don't care, and I don't need to make any sacrifices for you."

This is the attitude of the Western world to climate change too. Western development experts promote so-called sustainable development programs in poor countries, proposing radical changes to major economic sectors, without any discussion about equivalent changes in Western societies. The Norwegians spent hundreds of millions of dollars on programs to stop deforestation in Brazil, Indonesia, Colombia, Guyana, Ethiopia, Liberia, Peru, Tanzania, Mexico, Vietnam and the Congo Basin instead of cutting their oil production at home. "You change your lifestyle, over there, in the poor world," this says, "we are more developed. We do not need to make sacrifices or change what we are doing." It suggests that the freedom of some societies, and some people, is more important than the well-being of the majority. This thinking must change.

Often, the way people think, or act, depends on a feeling that someone has their back. Individuals put themselves first when they feel nobody is looking out for them, or when employers or authorities seem to not be "on their side." In a world where your employer is trying to grind every extra hour out of you, where you feel like you're in competition with everyone around you for the good things in life, to get ahead, for a little time and space and dignity, it quickly becomes natural to look out for yourself and think of your own needs first.

But when a person is confident in the community around them, there is less need to act selfishly. People don't have to closely guard what they have in a community if there is as much "give" as "take," and where the gap between you and your peers or neighbors is small. Humans are not islands. People do not have to struggle alone. Integrated, cooperative societies, where people care for each other, are happier, with fewer mental health problems. What's not to like about that? Isn't that something to aspire to?

How are we doing?

There is another side to the individual pursuit of pleasure and material possessions, and it is important to keep in mind if the new system is to endure.

People need to feel that they are doing well, and making progress. They need to feel that they are rewarded for their efforts. Status symbols like a nice car, an expensive phone, and fancy clothes make people feel good compared to their peers. It is natural for people to compare themselves to others. But this has also been manipulated by corporate propaganda and social media in recent decades, and this has taught people to believe that success, beauty, and desirability are reflected in what they buy.

This is something the creators of the new system can learn from too. People can be encouraged to rethink their aspirations and behavioral norms through strategic use of marketing, social media, branding, positioning, and word-of-mouth. (See Adbusters, for example.) The propaganda techniques that have encouraged overconsumption, a "me first" attitude, and which made "more" seem logical and desirable, can be used to encourage other values. New strategies can help people understand that cooperative behavior is not "only" needed to save the planet and secure humanity's future: it can also bring greater personal fulfillment.

In their early response to Covid-19 it was clear that many countries did not understand what was needed, or why. Some government leaders even claimed that the virus was just like a "flu." They thought that only the elderly and infirm could get really ill from it, although this also showed the lack of care for these people.

In truth, of course, the virus is very dangerous to many people. It spreads easily from person to person, traveling on breathable water droplets in the air. It survives on surfaces for many days. Those who are infected can spread the virus for weeks and not show any symptoms.

Because some people in government did not listen to what the scientists told them, because it was not what they wanted to hear, the message from political leaders was confused in some countries, such as the US and UK. So people could not see why it was necessary to make changes to the way they lived. In countries where the risks were clearly and consistently communicated, and there was a greater sense of community, people understood the dangers, and made sacrifices.

In the same way, many societies do not understand what is needed to stop climate change, or why. They do not know how urgent it is, or understand the scale of what is happening. They think that recycling, shopping responsibly, and investments in wind farms will fix the problem. They think that it will be possible to make a transition away from fossil energy over decades, and that there is no need for radical change now.

New thinkers and communications experts need to help these people understand how quickly the climate problem will get out of control. They need to understand that only if everybody makes sacrifices now, together, can a terrible future be avoided. Only if people truly understand, can they really care. (If they really understand the climate problem and still don't care they are psychopathic, which is a different problem.)

Stinking colonial thinking

Apart from individualism, there is another major cultural attitude, especially prevalent in European and North American societies, which is also a barrier to change. This is closely related to the "we're more important" attitude that spawns double standards on climate action, but it is reinforced with weapons that kill.

In the eighteenth and nineteenth centuries, most of Europe and the UK got rich by building colonies – let's not be fancy here: a more honest explanation is "by stealing." In those days, it seems, it was socially and politically acceptable to sail your ships to someone else's country and say "hello, these are our guns and cannons and, by the way, all your wealth and people belong to us now."[22]

These days, most people think that's no longer the way things are done, even if they aren't really aware of how deep that history of rape, plunder, and exploitation goes. Most people, if they think of those times at all, believe that imperial dominance, theft, and resource-grabbing at the point of a gun is in the past.

Unfortunately, it is not. It is still alive and well, only today it has a different name: these days it is called "free trade" and "good business." The link between then and now, between colonialism and modern business, looks like this:

Before	**Now**
Western countries believed their people were superior, modern and advanced. God had blessed these countries with technology and superior knowledge.	Western countries believe their businesses are superior, modern and more advanced. These countries are "developed" because science has given them technology and superior knowledge.

Western countries believed that other, "less-civilized" countries should learn to be more like them. They should follow the same God and allow Western countries to extract and export their natural resources using local slave labor.

Western countries believe other, "less-developed" countries should learn to be more like them. They should adopt the same economic ideology and allow Western companies to establish manufacturing sites where labor costs and environmental standards are lower, and still export their natural resources.

Western countries believed poverty in poor countries was because local people were undisciplined and had not learned to become civilized. They did not think it was because Western countries were extracting all the wealth and keeping poor countries stuck in a rut.

Western countries believe poverty and environmental degradation in "less-developed" countries is because local people are undisciplined and have not learned to become sustainable. They do not think it is because Western countries are extracting resources, demanding cheap manufacturing, and keeping "less-developed" countries stuck in a rut.

Post-colonial thinking is just a rebranded version of colonial thinking and is a major barrier to change. Western countries dominate trade and diplomacy around the world. They decide the rules of global economic participation. It is impossible for the now "liberated" ex-colonies to build sustainable economies, because they are forced to provide the manufacturing sites, cheap labor, and natural resources that the economically dominant countries demand. Western countries have, for centuries, cultivated the elites in the poor world, and encouraged them to sell off their nations' wealth. This strategy was the key to Western power, from India to Indonesia. No wonder countries like China see no moral contradiction in doing the same today.

By insisting on open markers, and by using finance and competition to restrict the technology and scale that companies in poor countries are able to develop, the rich world keeps the poor world permanently poor. The countries of the poor world can never do anything more than mine, chop, and log unless they put up trade barriers to protect their companies as they grow. But that is impossible without incurring economically crippling trade sanctions. Just as in the nineteenth century, the poor world is trapped in a perpetual half-life, dancing to tunes chosen by people on Wall Street and the City of London. The only countries to have successfully got around this problem are Japan, South Korea, China and, to a slightly lesser extent, Vietnam and Thailand. These countries put up trade barriers, though they were often hard to spot, promoted their own national champions for many years and then gradually opened to the rest of the world when they were ready.

In the past, the consequences for many countries which tried to keep their resources, upscale their economies, and become self-sufficient, outside that Western-dominated system, were clear and bloody. Indonesia, Panama, Congo, Venezuela, Cuba, Iran, Chile, Iraq – countries which tried to control their own destinies had their leaders removed (or assassinated) and replaced with a Western-friendly government, so the resource plunder could continue. Sometimes the job was done at the request of private companies, as in Guatemala on behalf of the United Fruit Company (now Chiquita).

It is exactly the same now. Western economic and military supremacy forces weaker countries to open their doors, just as they did in the nineteenth century. Today's gunship "diplomacy" has an additional twist, however. Western countries are also now able to make profit through arms sales to the countries which have been economically colonized. The US, France, Germany, and the UK are among the world's top weapons exporters.

And, as an important aside, why exactly is the rich world continuing to focus so much money and effort on an industry which is designed to increase death, misery, and division? Who can sleep at night, investing in murder? Why is so much creative and technological energy put into killing, torture, and pain when this creativity and these funds are needed to save the planet from runaway climate change?

What if all that energy and money went into building better communities instead of blowing them apart?

Chapter 11

Where should the transition begin?

We have talked about what a new system needs to look like, and how a political movement will be needed to make it happen, and we have discussed some of the hurdles which will need to be overcome. A further question relates to where the transition can begin and where people can be found to dedicate themselves to it.

As before, we have looked for examples around the world to offer inspiration.

When it comes to implementing the social changes needed to address the climate problem, some countries stand out more than others. In Germany, Austria, Scandinavia, and the Pacific island nations most affected by rising temperatures and sea levels, understanding about the scale of change needed is generally good. The majority of people in these countries see what needs to happen and many are willing to act collectively in society's long-term interests.

In several other countries, there are often some very large pockets of understanding but there is also very strong opposition from finance, industry, and climate deniers. That makes it harder to build the political consensus needed for change, though it may be possible in some particular cities, regions, or states. Here, we are referring specifically to the United States, Canada, the UK, and Australia.

There are also countries where there are small pockets of understanding, such as France, Spain, and Italy, and there is China where there is a good level of understanding in the government about the threat of climate change, but a poor level of understanding among the people. China is the world's worst polluter, by far, and yet it has paradoxically also claimed that it wants to build "an ecological civilization." This is at odds with their development model over the last 30 years, which has been almost entirely based on rapid economic growth, very high industrial production, and massive resource extraction, with terrible environmental consequences. Since the onset of Covid-19 the country has also shifted away from renewables and back to coal, making the climate problem worse.

In some ways, this is puzzling, because the Chinese government knows that hundreds of millions of its people will be very badly affected by climate change, and that cities like Shanghai, Shenzhen, and other parts of the Pearl River Delta will be uninhabitable[23] before 2050 because of rising sea levels.

There are countries where there is outright hostility to any discussion of climate change and fierce opposition to any response. These are mostly countries that are highly dependent on the fossil fuel industry – in the Middle East, Russia, Central Asia, and the Caucuses. This is understandable. Without fossil energy, many of these countries would be very much poorer than today. They will need careful support during the transition so that they are not made worse off.

There are also a very large number of countries where understanding of climate change is low, both within governments and among citizens. We come across intelligent people in these countries all the time who still think that climate change is something to do with the ozone hole. These countries are mostly in the poor and middle-income world, though we would also include Japan, South Korea, and Taiwan in that group, places with very well-educated populations. This group could be especially interesting because they are more open to learning about what is happening. Many citizens of these countries are used to thinking long-term and working with others for the good of all.

As well as nations, there are institutions which could supply people to work on the transition or act as catalysts for change.

Some institutions understand the situation very well and can play a positive role in the transition, including the Catholic church, as well as some other religions, the United Nations, and many military organizations around the world. The German and US military are particularly well informed about climate science because they see, rightly, that it presents a major threat to social stability and raises the risk of armed conflict as people fight over land and water. They know too that one of the big short-term consequences of climate change will be a huge rise in migration, driven by higher temperatures and more prolonged droughts. They understand that any meaningful response to climate change is closely tied to the fate of the fossil fuel industry.

There are also institutions which are mostly blinkered to the climate problem because, for years, their work has depended on spreading a neoliberal worldview. This group includes the World Bank, the World Economic Forum, the IMF, and the European Central Bank, institutions which have promoted economic growth and free trade. This is not to discredit individuals within those institutions who have dedicated their working lives to creating positive change, or who try to create change from within. There are many who, also in retirement, continue to question and refine their personal choices and try to apply their experience in positive ways. However, the backbone of these institutions has been free market economics and the promotion of economic growth.

Some of these institutions have belatedly tried to change their message, even producing reports which call for change. Yet their intellectual challenges remain hard to resolve because for decades they have enthusiastically promoted the system that is the cause of the crisis today.

We would also include parts of the media in this group, including magazines like *The Economist*, and newspapers like *The Financial Times* and the *Wall Street Journal*. They too have almost unquestioningly promoted a system of human development which creates ecological destruction and widens inequality.

Another institutional group that is likely to be less helpful than it might initially appear is the NGO sector, especially some of those organizations devoted to green causes. While these institutions are often populated by nice, well-meaning people, there are only a small number that fully understand the enormity of the climate problem, certainly in our experience. The people working in these institutions care, and often very deeply. But those that focus on only one part of the problem (species loss, water pollution, social injustice) do not always see the whole.

Some NGOs have even become part of the establishment and so shy away from challenging the system. This is partly because of the way they are funded, though some have also been seduced by neoliberal thinking. They have been persuaded that the free market and the finance sector can solve the world's environmental crises, blinding themselves to the unfortunate reality that they are the cause.

Within these myriad groups and nations there are people who can help manage the process of change. The trick is to find them. So, this needs to be one of the other early tasks for change-makers. Identify champions, communicate with them, and build networks around those who really get it.

Finally, in this section, and just for the record – we are asked, surprisingly frequently, if humanity should simply move to another planet. So let's be clear: that's not possible. Even if societies were able to launch a spaceship that could carry thousands of people for many successive generations, it would take many thousands of years to reach what astronomers think is the next potentially inhabitable solar system using the best currently available technology. Even then, as scientists do not fully understand how to stop a spaceship traveling at nearly 650,000 kilometers an hour in a vacuum, it would simply sail past its intended destination.

Chapter 12

What can I do as an individual?

There is so much you can do. People acting *alone* cannot change the outlook to any useful degree. But *collective* action on a global scale, can have an enormous impact. This section contains lots of essential, practical advice on how you can foster change, acting collectively.

Communicate, communicate, communicate

Changing the way societies view the world, how they think about economic growth, individuality, democracy, climate change, and species loss, will take a phenomenal amount of work. Change-makers need to be persuasive and have excellent communication skills. Encouraging people to change the way they think takes time, skill, and patience. It is also complicated because it needs to take place on so many different levels. On the surface, it might seem that people quickly "get it" after an evening seminar, or a three-day workshop. But deeper habits of thinking are much more powerful than those that are front-of-mind.

Exercises to persuade people to see the world differently need to work at the surface level and at the deeper, subliminal, "muscle-memory" level where reflexes and habits lurk, to reinforce and embed the message. Take dieting as an example. Let's assume I tell myself that I need to drop 10lbs (4.5kg). I plan for week-day lunches of steamed vegetables and tofu. I go to the gym. I deny myself ice cream. My mind is convinced that my plan will work. But within a few days, one mid-afternoon, my hand starts wandering towards the cookie jar and, without noticing, my arm, legs, and feet follow. Some deeper force inside remembers how delicious those chocolate cookies are, and how good I feel when I eat them. Front of mind is clashing with deeper, subliminal forces.

Try this one: you've lived in the same house, in the same street, for 15 years. Then you move to another part of town. For weeks, planning, logistics, and the everyday realities of the move fill your consciousness like sand in a jar. There is no part of you that doesn't know you are moving house. But three weeks later, you get in the car to drive home from work and you suddenly find yourself driving to your old house. Some deeper memory is still there, directing your hands on the wheel and your route-planning mind.

It takes repetition and reinforcement to change thinking.

1. Learn the Facts

First, you need to learn the facts as well as the logic of what needs to change and why, until you know it backwards. You need to learn how to present a case and give yourself a crash course in persuasive communication. Always speak from the heart but with a rock-solid foundation of fact. If you know your facts, and understand the science and the arguments, you will be better prepared to respond to awkward questions.

Learn the facts like you're going into an exam. Imagine what the exam questions might be. What about the trick questions designed to throw you off? This is a bizarre world where facts have become optional extras, where winging it and cheating have become as common as using one credit card to pay off another. But physics does not defer debt to a future which never comes. You can't put a lid on a volcano. Climate change is about facts. Nature is not even the teeniest bit interested in fancy accounting, statistical manipulation, and lies. The planet is not persuaded by sleight of hand. Now is the time to respect facts and the truth. As Malcom X said: "Truth is on the side of the oppressed today, it's against the oppressor. You don't need anything else."[24]

2. *Speak up*

Then go and speak to your parents, your neighbors, and all your friends, and test your communications skills. Your networks are the best place to start. Communicate the urgency, explain the facts, and take note of their responses. If someone is skeptical or dismissive, don't take it personally: conflict with those close to you will not help. Listen to their objections. Are they based on emotion, structural realities in their own lives, or a misunderstanding of the facts? Try to understand, because that helps us all to move forward. But remember: the objective is to communicate and persuade, not to compromise. There is no time for half-measures, or weak-hearted solutions.

Then ask those you convince to do the same – to speak to *their* parents, neighbors, friends, to spread the message. Understand, however, that some people will never get the message. It is not possible to convince everyone. Based on our experience, 20–30% of people in the rich world have a relatively good understanding. They know that recycling is not enough. Another 20–30% are stubborn and refuse to listen. They will choose to sing the la-la song no matter how well-communicated, logical, or persuasive the message you deliver. Those people are not able, psychologically, to internalize the truth. Perhaps it is just too hard to accept the pain of knowing what will happen, or perhaps it is too painful for them to accept their role in adding to the problem. Perhaps they just enjoy the rewards of their current life too much to give it up. Sometimes the people who cannot be convinced will be strangers. Sometimes they will be friends or family. It may be hard to accept that someone you love is never going to get it. Keep trying, but don't lose yourself in trying fruitlessly to convince someone who cannot listen.

3. *Toughen up*

The push for change will inevitably lead people into conflict with others, so change-makers (us included) need to toughen up emotionally, in order to keep going. Sensitivity is clearly a good thing: the new system must have empathy at its core. But as we discuss shortly, there are many emotional blows ahead and (believe us: we know) they can flatten the fight right out of you. Try to turn the empathy into something constructive, to try and understand more about the challenge.

You don't have to change who you are, but there are ways to make the pain you feel useful. Toughening up doesn't mean you have to grow callouses on your heart, or somehow learn to not feel – that's damaging in other ways. Rather, it's about strengthening your resolve to turn your pain into something good. It's important to acknowledge the pain, to call it by its name, and not to deny the terrible things that are happening.

Those who can respond positively to their emotions build strength. Those who have experienced climate grief have a valuable ability: so talk about it. Talk to others about how it hurts and why. And then share it, widely. What's the point of hiding away? You're not going to save others from suffering by suffering alone! There is more about this in the next section.

If you build relationships with like-minded people it is easier. Work with the 20–30% who already get it, to persuade the 40–60% who are open-minded but ill-informed. But remember too, that no matter how tough people are, the reality of what is happening will still be painful for them to understand.

4. Build your network

Building a network so that you are part of a like-minded community brings comfort and strength to those suffering from climate grief and the sense of isolationism which often accompanies climate action. There is a very powerful positive emotional shift as people learn to view themselves not as individuals, alone in a sea of troubles, but as an integral part of a movement that is dedicated to shared mutual responsibility and that cares about wider society. Work with those who understand this too. That will help you stay strong. Check in regularly on the information links at the back of this book to remind yourself about what is happening and make new connections. Set up regular meetings, book clubs, and support networks for those who understand.

5. Keep the goal in mind

Remember that the goal is to create a better world, with compassion and dignity. Cynics may tell you that what you are trying to achieve is Utopian nonsense, that it's all very well to dream of paradise filled with peace and love, but what really makes the world go round is money, power, and knocking a few heads together every now and then to keep everyone in their place. Cynics may tell you that bad apples will always cause trouble, that it takes more than peace and love to keep societies running. Well, that's probably true. But starting from a principle of dignity, equality, companionship, and compassion is surely a better place to start than a base of misery, inequality, loneliness, and greed. Remember the goal.

Be prepared for push back

The front line of change-making can be an exciting place, but it can also be lonely at times. Especially at first, you may feel like the lone party-pooper crying "hurricane!" while everyone else is still partying on the beach. The most beautiful weather often precedes a storm. But this also makes it hard for many people to imagine there's even a little shower coming, let alone apocalyptic change.

People may tell you you're crazy, or they may just ignore you and party on. They may reason that if it were as serious as you say, people in authority would have done something, said something, put out a warning by now. And who are you anyway? You're not some kind of government-backed climate scientist! How come you're so smart and everyone else is so dumb? People will likely roll their eyes, and mock you for taking this seriously.

Knowing about the reality, the immediacy, the consequences of climate change while everyone else keeps partying can make you feel like you're in some terrible dream, watching puppies scampering on a busy freeway with cars and trucks driving pell-mell around them. You are screaming for drivers to stop but no sound comes out. You might feel like you are going crazy. You will probably doubt yourself, and doubt that this is really serious. You may start to wonder if everyone else is right. People may tell you that all this climate stuff is a hoax. Take comfort: you are in great company. It is very common for societies to exclude the people who speak out about climate change. Greta Thunberg, the young Swedish climate activist with the blonde pigtails who told all the suits at Davos they should be ashamed of themselves, is mocked and slandered all the time. People have said she is mentally ill, an indoctrinated tool of her parents, irritating, and an annoying little brat.

We ourselves are well aware from personal experience that even those who are supposedly on the same side can work against you. A couple of years ago three journalists from Greenpeace Magazine came to interview one of us. They came into our home in Zurich, asked about something we had written, our activism, and our background and then (here's the really clever part) published an article which poked fun at us for using coasters to protect an old table.

That's just weird.

Ancient tropical rainforests are burning down (releasing all their stored carbon into the atmosphere), encouraged by giant multinational corporations, and Greenpeace chose to mock someone fighting for change for respecting a piece of furniture. Massive fossil fuel companies with more money than some medium-sized countries have been lying about the hydrocarbon-climate change link (in fact, actively denying the link) for decades. Yet Greenpeace's eco-warriors decided to denounce someone on their own team for recycling. Way to go.

Also common in the fight for radical reform is something called "gaslighting,"[25] a particularly sinister, corrosive, and abusive form of psychological isolation. Gaslighting is the systematic erosion of another person's connection to reality by denying facts, their physical surroundings, or their emotions or perceptions.

If an individual tries to correct damaging behavior, and the other person simply denies that it is even happening, that is gaslighting. You may have experienced it in "playful" forms, where someone close to you denies having done something you both know they did, but so convincingly that you actually begin to doubt what is true. Hard-core denial by some politicians about climate change is becoming less common, but the link between fossil-fuel combustion and changing climate is still fair game in some countries. Scott Morrison, who is now the Australian Prime Minister, once waved a lump of coal in the nation's parliament, goading members not to be afraid of it. Fossil fuel companies have practiced gaslighting like this for years. It's deliberate. It's a classic form of the ploy to divide and conquer, to keep the opposition divided, confused, and so unable to organize itself to challenge authority.

If you are faced with this, it may be useful to recall the Gary Larson "Far Side" cartoon produced around the time of Bovine Spongiform Encephalopathy (BSE or mad cow disease[26]) in the 1990s in Britain. It pictured a cow lying on a psychiatrist's couch saying "maybe it's not me that's mad, you know? Maybe it's the rest of the herd that's insane..."

It may be little comfort sometimes, but the truth with climate change is that the rest of the herd really *has* gone insane. Think about it: to experience record temperatures year after year, increasing storms, droughts, and massively more wildfires – and all this happening in places where you live, where your friends live, where you vacation – to see all this with your eyes and experience it with your body and *not* want to intervene and somehow stop that happening, you really do have to be insane. The only insane thing is to *not* react.

So when someone tells you to stop worrying about what's happening or tells you to get a proper job, or have children so you wouldn't have time to worry about all this, take a deep breath. People often attack others when they feel threatened, or frightened, or defensive about their own life choices. Societies find it hard to accommodate differences from the "normal" way of thinking, and they resist any push for change. Normal may not be good, but it's easier than change. Agitators for change are *annoying* to those who want life to be easy.

Remember: you are not alone. You are the sane one!

Be prepared to grieve as your eyes learn to see again

When you really understand what is happening, it is hard. It is like a bereavement. This is why so many climate activists suffer from depression. It can feel like standing in the middle of a river, trying to stop the raging water with your bare hands. The grief can be searing because there is so much to grieve about. The destruction of so many beautiful, sophisticated ecosystems, crushed under shopping malls or destroyed by mining, forest fires, and rising sea levels. The death of so many innocent creatures: the mountain marmots and the Indonesian proboscis monkey with his magnificent long nose, whose forest home has been burned and razed to the ground just so some financial speculators in London or Frankfurt or Hong Kong can make a tidy profit in palm oil futures.

Humanity could have avoided all this. Yet those in charge have chosen to do nothing, even though they knew what would happen. They knew about climate change, and they knew about the suffering that would result if they did not act. Powerful, influential people have known about climate change for years, but have deliberately concealed and twisted information to prevent change for their own selfish interests. Who, with knowledge and a conscience, would *not* grieve at the greed of those in big business, as well as those who finance and support them, when they still lie, even now, to protect their money?

Grief and anger are a natural, compassionate response to such outrage. But those of us who are fighting for change must stay strong and gain strength from each other. Seek out others to share what you feel and try to help others who may be struggling. People cope or struggle in different ways, so remember to check in, call, and listen, with compassion.

Remember: you are not alone. There are things each of us can do. But divided, we fall; united we rise.

Bust those myths

One other very important area where you can make a difference is in myth-busting. Many people mistakenly believe that they are genuinely helping to avert planetary disaster by recycling, taking fewer flights, buying an electric vehicle, or donating money to charities. This is also a major barrier to change.

People have to realize that *none* of these things does any good on their own: in fact, they often make things worse, because they promote a false sense of progress. Because many people think these actions work, as a global society, humanity does *less*.

Myth-busting is essential.

A good place to start is with many people's messianic obsession with technology. Most conversations about climate change turn pretty quickly to the magic fairy dust of technology, or to clever financing. But anything that allows the central elements of the system to continue as now, endlessly extracting more resources and producing ever-more emissions, cannot work. This is a difficult reality for many people to grasp, especially those who have been indoctrinated into the dangerous belief that humans are smarter than nature.

People will tell you that technology and science are great. They are the reason people don't die aged 35 from typhus or the Black Death any more. It is thanks to human innovation that societies have clean drinking water (well… some do), SUVs, and diabetes medication, so they can live the lives as they want instead of being slaves to nature.

But this *masters-of-the-universe* thinking encourages societies to think that human ingenuity, rather than behavior change, can solve every problem.

For centuries, alchemists tried to turn base metal into gold. Now they know it is possible. All that is needed is a particle accelerator and huge amounts of energy, though the amount of gold produced is tiny. Thanks to technology it is also possible to grow diamonds without waiting millions of years. Scientists have learned how to split the atom, clone sheep, and grow meat in a laboratory. It is also possible to seed clouds to make rain.

These are all astonishing achievements and, to many, they are proof that humans can triumph over nature. But these scientific developments, impressive though they appear, are really only proof that humanity can tinker at the edges of a vast, finely-tuned, awesomely complex physical, chemical, and interconnected biological system.

The planet's atmosphere is changing chemically, and on a scale that few people can even imagine. Slowing this process is about the natural laws of chemistry, biology, and physics. There is no technology that can stop climate change, or control the uncountable heat waves, droughts, floods, storms, burning forests, melting permafrost or the multiplying viruses, insects,[27] and bacteria that will result. Repeat after us: *Technology. Will. Not. Save. Us.*

But people, working together, with compassion and a dose of reality, just might.

The best chance humanity has is for people to work together for change. That might sound crazy. But it's less crazy than thinking some geeks in a garage are going to come up with a techno-fix that will make them millionaires and save the planet in the nick of time.

Plan and protect

In developing the blueprint for a better future, it will be tempting to bring in experts from the old system. This is a weak point in any revolution, and should be avoided: the old system is toppled, and new people take power. But while the cheers of celebration are still reverberating, the overseers of the new order start to see something: empty spaces, vacuums of power and skill and expertise. Skilled people are needed to guide communications, social welfare, agriculture, diplomacy, economic policy, to get the wheels of society turning again. So, the new order brings in some experts from the old order to fill the vacuum. And in comes the old thinking, the old ideas, the unreformed assumptions and approaches and priorities. And that's when the revolution fails.

This is a hard trap to avoid. In an ideal world, and we are not in one of those, the leaders and technocratic advisers of tomorrow would be trained in advance, so that they were ready when circumstances demanded. That's not possible. But those fighting for a change in government can still think carefully, in advance, about how to approach this. They can try to identify skilled individuals, with a deep understanding of the climate problem and its causes, who can help train the leaders and technocratic advisers of the future. They need to approach the challenge as if they were building Apollo 11, which was also constructed without the real conditions in which it would operate. As with the space missions of the 1960s, there is a lot that can be planned in advance.

Key skills will be needed to build a new economic system, based on providing essential products and services including food, medical care, and transportation. Jobs will be needed to employ the hundreds of millions of people whose livelihoods have been lost. They will need to be retrained and supported financially and emotionally as they make the transition to new jobs and workplaces, with new and unfamiliar colleagues and expectations.

Behavioral psychologists will be needed to help guide, acknowledge, and manage the wide range of emotions that will emerge across society during the transition. Skills will be needed to build a new social system so communities can provide pastoral care to one another, with support groups to help those who are struggling with the changes. There will be a lot of anger too, for example, from people who don't like hearing that they can't fly to other countries for their vacation, or who resent the changes.

This is likely to be more challenging in the US, where the idea that people should be mostly free from government control is very deeply rooted in national thinking.

In almost every society, skills will be needed to rethink the role of politics and political institutions, as well as finance and banking. Many of the structures and practices of modern political systems are legacies of old ways of thinking. The new system will need to redefine the social structures necessary to achieve the primary goal: preventing runaway climate change by creating a stable, ecologically-sustainable global society. The long-term role of money and finance will need to be carefully rethought too, as we have discussed. It will be easy to fall into the trap of old thinking.

New thinkers will need to constantly ask themselves, and each other: "Why? What does this achieve?" It is hard to ask these fundamental questions about topics societies have been taught to take for granted, but it is precisely the structures and habits of life that are *not* normally questioned, that need most urgently to be questioned. And beware: many people who do not want change in these areas will make convincing sounding arguments against change, to frustrate those building the new system. It will take intellectual rigor, a clear vision, and a strong, honest heart to weed out the self-serving, greedy, manipulative agendas. Do a lot of research before accepting advice from anyone: look into their education, their career background. If they were previously connected with the kind of thinking that has created this mess, be very cautious. They will need to prove how they have reformed. Do not easily be persuaded, even if they seem convincing.

In designing policies, strategies, and structures for the new system, don't forget that different communities have different realities. Their experience of change will be different, and that means they will likely react differently. A poor community with shoddy infrastructure and few job opportunities may react with anger to being told that they have to make sacrifices.

Millions of people have been marginalized economically, politically, judicially, and socially by the existing system. They have watched opportunities, money, privilege, and equity circulate in middle- and upper-class communities but not their own. Change-makers will need to do their research and tread with sensitivity and respect, before asking those who have been badly treated by the current system, to give up anything they might see as security, dignity, or comfort, even when this feeling might seem irrational.

Of course, theorizing is different from doing: Covid-19 demonstrated that. The US and UK were said to be "the best prepared" to handle a pandemic, according to a report[28] presciently published in late 2019. The report assessed the capacities and capabilities of countries to respond to a biological threat. Apparently though, the authors assumed that these would operate in a vacuum, isolated from the conflicting priorities, personal agendas, and irrational dislikes of reality. Everything that is unavoidably, inherently *human*, did not come into their assessment of how pandemic planning would work.

This is typical of many sustainable development programs

too. Time after time, those who design sustainability programs, forget to think about how people react in real life. Humans don't do as they *should*: they do as they *do*. So plan for reality, and think about the way people actually behave. Think about the stupid, instinctive, messy, and annoying ways people behave, and build that into your plans.

And beware of using "culture change" programs which are often infected with the standardizing disease which totally ignores the realities of human behavior and the natural world. Humans are governed by friendship, fear, greed, hunger, love, jealousy, tiredness, lust, and boredom – all the things that look wishy-washy and unscientific in a managerial report, psycho-babble that no sensible person should take seriously. But these floppy concepts are also the issues that make clever, glossy managerial change programs fail.

Societies only know how they will react to change when real life plays out. That said, planning ahead and training in key skills can help reduce the shock of change when it comes. Many obstacles and risk factors are predictable, as long as planners think like humans and not like sterile automatons. Learn to ask the right questions. Plan for people, not instant-noodle Utopia.

Organize and push

Once you have spoken to all your family, neighbors, and friends, and armed with the fact sheet at the back of this book, push for a meeting with the high-level decision-makers in your organization: the dean of your university, the head teacher of your school, the Vice President, General Manager, or CEO of your company. Tell them:

This is what is going to happen

Humanity has less than 15 years to stop a climate change chain reaction and needs to cut its greenhouse gas emissions by 60% in the next decade.[29] If it fails, hundreds of millions of people will die. Everybody has to change, including the people who work or study here [insert the name of your organization]. If not, this organization will be contributing to the deaths of hundreds of millions of people. We want to help change how the company/school/university behaves and do the right thing.

This is what the organization/academic institution needs to do

- Sell investments in companies with high greenhouse gas emissions (such as fossil fuel producers, airlines, car manufacturers, cement, or shipping), which extract non-renewable resources, or which damage ecosystems (such as mining). Ask the organization or school to publicize news of its divestments.
- For schools/universities: In *every* academic discipline, demand climate change be added to the curriculum even if it seems minimally relevant. For example: The law faculty should not teach students to protect companies that damage the environment; economics, management studies, and finance should not teach societies to always prioritize profit, and ignore environmental consequences; accountancy should measure the destruction of nature; architects should only design buildings that need little or no energy.
- For companies: Show how the products or services damage the climate. This should be in absolute terms, not comparative (e.g., how a product or service is less damaging than others, or than previous generations of products/services).
- For schools and companies: Demand a draft plan showing what the organization will do to change.

Companies that directly produce or use carbon-intensive products or services should be encouraged to stop. Some might be able to shift to alternative products or services. Start the ball rolling about what is possible and fuel the debate. Help the company's leaders understand that there is no other way, and that they can seek support for their employees if they wind up the company. Build the pressure from inside. We understand, of course, that this will simply not fly in some companies.

Communicate across your network about the action you have taken and how your organization responds. Learn from others, and help others learn from your experiences to refine what they do too.

Activate and take down

For those prepared to take a more activist role, fasten your seat belts: set your sights on the activities that are blatantly, unapologetically producing climate-heating emissions, the ones that don't even pretend not to, and don't try to hide. They're quite easy to spot when you start looking. They're the ones that had the excuse of ignorance 50 years ago to behave without a care for the future, but whose continued activities are inexcusable today. Take, for example, Formula 1. Great fun: a bunch of (mostly) guys racing very fast cars around a track each week. Thrilling, high-octane stuff in the 1970s, with all the crashes and bad boys and speed.

But in a world rapidly approaching a catastrophic tipping point, what possible purpose does this circus serve? Has anyone in F1 (the equivalent in the US could be NASCAR) even thought about how callous, how disgustingly sadistic the whole activity has become? These people are celebrating the combustion engine while the planet burns! That's like setting up a barbecue competition in communities destroyed by wildfires. What the hell do they think they're doing? How dare they? Here's what needs to happen:

- Target every company that sponsors F1: Red Bull, Tag

Heuer, Rolex, Rauch (that's a fruit juice manufacturer. Why it wants to be associated with gas-murder we don't know), Exxon, Daimler, Aston Martin, Citrix, Rexona (it beats your BO but by supporting F1, it stinks), Martini, Dita (optical experts: shame they can't see), AT&T, Oris, Hackett, DHL (delivering Armageddon?), UPS, Heineken, JCB, Oris (counting down the seconds to unstoppable climate change), Acrons, IBM, UBS, Johnnie Walker, Zepter, Puma, Kasperski, Riva, Hublot. These products sponsor death.

- Put pressure on your friends, family, and neighbors as well as your organization to stop associating with these brands. If you drink Red Bull, if you stock Red Bull in your company's canteen, you are sponsoring murder.
- Any country or city that hosts F1 needs to go on a blacklist. If your city hosts F1, put pressure on the city council to pull out of this shameful, murderous activity.
- Push against coverage. Ask any media channel that covers them: "What are you doing?" No serious media operation would televise bear baiting or tobacco smoke-a-thons in a room full of children. Why are they televising F1? Push for a news blackout of motor racing.

All those famous racing car drivers, who are hailed as role models and champions, should be confronted with the stark truth of what they do: Mr Hamilton, Mr Vettel, Mr Ricciardo, you only have one job description from this moment on: "I burn hydrocarbons." You are not heroes. You have no honor. No small boys should hold you up as an icon of manhood. This is not the 1970s. You are not Niki Lauda or James Hunt, whose consciences were at least free of the stain of climate change. You know damn well the consequences of burning hydrocarbons. How can you get behind that wheel?

And don't stop there. If you can run a successful campaign

against one organization, once you've learned what works, sharpen your blade and choose the next. Through social media and your network agree on the target – an oil company, an airline, a cement producer or a car maker. But choose only one. Then do the same. Arrange to boycott its products or services, and have everyone you know boycott them too. Spread the word, write articles, and put up posters.

Those taking a more active route to change should also remember that flexibility is needed, such as that displayed by protesters in Hong Kong during the summer of 2019. They adopted a flash-mob approach to activism, emerging and then disappearing like vapor.

Most companies are vulnerable to this sort of activity, especially since the Covid-19 crisis began. A 10% drop in sales is hard for many companies to survive. So target a 25% decline, as a first step. Work with overseas networks too, so that your impact is global.

There is also nothing illegal about this. So work together to target the polluters, the rich, and those who have promoted fake news and climate change denial. Organize a boycott – avoid products, demonize and shame those who use them. But always focus on one company at a time so that your fire is tightly focused. Targeting lots of companies at the same time will have much less impact.

Getting the needle in

As well as boycotting those who are directly responsible for humanity's plight, those who seek change should look for other pressure points of influence. Street demonstrations are too easily ignored and risk arrest. So, try to identify points of influence that have a useful effect, a little like finding the parts on the human body for acupuncture needles. Demand changes that seem counter-intuitive but that also make sense, and which will make people sit up and pay attention. Think of using a million straws to break the system's back.

For example, lobby your political representative (at least until you have successfully replaced him or her) for them to regulate for a shorter working week. This may not be as simple in America as elsewhere, but even in the US, try to start a debate about vacation time, because American workers toil for longer than most of their peers. Regulating for more vacation time creates jobs. It also makes people happier because they don't have to work as much. Think about it this way: there is a fixed amount of paid work to do every year. If the people doing this work have shorter hours, employers have to hire people to make up the shortfall. When so many people have lost their jobs because of Covid-19, the need for societies to share the available work better is higher than it has been for decades. So push for a four day week. Who doesn't want that?

At the same time, push for other reforms which help move the system in a better direction or highlight some of its flaws. Demand that those who look after children or the elderly at home get paid by the state for what they do – and properly. Their work is extremely valuable to society, and they should get paid for it. To fund this, demand higher taxes on the rich and high death duties. Then everyone starts life with the same. These ideas might seem crazy – but they are also in the interests of most people and so they should gain majority support. Encourage people to think differently!

Finally, demand a reallocation of government budgets. Demand less spending on defense and more on combating what is the much greater threat to everyone's safety: climate change. Tell your representative that all the money that is currently wasted on state-sponsored murder should be devoted instead to responding to climate change, and tell them why. We know that this too will likely fall on deaf ears, certainly at first. But keep pushing, because a world without weapons is not only something to dream about. If no one pushes, it will never happen.

Chapter 13

Join us

At the start of part two, we asked you to imagine an army driven into retreat, closely pursued by its enemies. It had reached a wide, fast moving river. We said that this is just like where humanity sits when it comes to climate change.

Now imagine that army again. It is hunkered down now. Frustration is growing, impatience too. The river is flowing very fast, silently, and rising quickly. There is a virus spreading across the camp. It is an alarm call, a last chance to wake up, stand together, and act.

On the horizon, something small and bright appears. Hearts warm in anticipation, breaths are held, eyes squint in the glare. It is infinitesimally tiny. Even now, close up, it appears too small to carry them, this conduit as narrow as thin wire, barely microns across.

A voice speaks a language that is both ancient and of the future. It promises passage, though almost no one understands. Nearly all are welcome. This is the chance to get to the other side, though safe passage is not guaranteed.

And there is a condition. Some must be left behind. Weapons must be abandoned, as must all those who were responsible for bringing them here, all those who were in command. They led everyone to this dark, desperate place. It is long past the time for these people to fall upon their swords.

The journey will not be easy. This is repeatedly made clear. There is so much that can go wrong, and many mistakes will surely be made. But the crossing is possible and it is the only way.

Yet there is uncertainty, still.

Thankfully, there are other choices to be made, and many are very good ones. There is the choice of a new, brighter future on the other side, built on dignity, respect, and care. There is the choice of a future that preserves the beauty and the potential of life. There is the choice to leave this place behind. This dank place where everybody is abandoned to slide through the mud, ever faster, into struggle, and fear, and hate.

Against all odds, there is this one last, wonderful, chance to cross.

If they make it to the other side, there will be new obstacles to overcome. But the vehicles and guns will no longer belch their hot fumes into the skies and there will be the opportunity to live in peace and prosperity, in harmony with the world.

Join us on this journey, this crossing to a better place. Let's prove to ourselves and each other that the passage can be made.

Every day offers each of us a chance.

Join us.

Chapter 14

Fact Sheet

The average temperature of the planet is rising, and the pace is accelerating. This is not natural.

- The warming is mostly down to the way humans produce **energy** and **food**.
- Average global surface temperatures are now 1.1ºC, compared to 1800, which is higher than at any time in the last 3 million years.

Impact:

- Mountains are crumbling as the ice that holds them together melts.
- Glaciers are disappearing and forests are dying.
- Storms are becoming more frequent.
- The number and extent of wildfires is rising.
- Droughts are becoming more prolonged.
- Some crop yields are already declining.
- Rivers and lakes are drying out from evaporation and too little rain.
- The permafrost in Canada and Siberia is melting, releasing more greenhouse gases.
- As the ices melt, less heat is reflected into space, increasing the warming.
- The most important GHG is **Carbon Dioxide (CO2)**.
- Before the industrial revolution the concentration of CO2 in the atmosphere was 280 ppm (**parts per million**).
- It is now 50% higher, at 416 ppm, and growing by 3 ppm a year, exponentially.
- The **tipping point** that societies have to avoid, when a

chain reaction starts, is **450 ppm**.

- That is in 15 years (as at 2020).
- The last time it was at that level was 45 million years ago.

If this happens:

- The warming will be **impossible to control**.
- The great forests around the world will die and the ice at the poles will melt much faster.
- Glaciers will disappear and coral reefs will die.
- By the middle of this century the average temperature will have reached its highest level in 10 million years. Many parts of the planet will become uninhabitable after 2050.
- Most of the planet will become uninhabitable long-term.
- This is also what will happen if all of the conditions of the 2015 Paris Climate Accord are met.
- Even if hundreds of millions of people chose to live 100% sustainably tomorrow, it would not be enough to stop the chain reaction starting.

Acting responsibly as individuals is not enough:

- If you cut your personal contribution to the atmospheric pollution each year to zero, starting tomorrow, you will delay the chain reaction by a fifth of a second.
- If everyone in America – all 330m people – stopped generating GHGs completely this would delay the disaster by two years.
- Action on climate change must include North America, Europe, Australia, Japan, India, China, and Russia.
- **Almost everyone** must cut GHG emissions by at least 7% **a year**.
- 20% fewer cars **within three years**, 20% fewer planes, 20% fewer coal fire powered stations and 20% fewer ships.
- In the following three years after that, another 20% reduction.

- The longer societies take to begin, the steeper the cuts have to be.
- GHG emissions must be at least **60% lower in 2030 compared to today**.
- By **2040** they need to be **zero**.
- Societies need to transform the way they grow food, and stop all deforestation.
- They also need to build thousands of **carbon capture plants** to bring the C02 concentration in the atmosphere back down to safe levels.
- If all this is done, there is a 50:50 chance of avoiding that chain reaction.
- Emissions **cannot be offset** in some way.

Almost everyone on Earth needs to urgently change the way they live, whether they want to or not.

Chapter 15

Suggestions for further reading and insights

Guerrilla Warfare, Che Guevara's 1961 book.

On Guerrilla Warfare, Mao Tse-Tung's book written in 1937.

Non-Violent Engagement, Popovic et al. This will help you think the battle through and understand what has worked in the past.

The Right Livelihood Way – A source book for Changemakers: 8 myths about non-violent activism (from a movement that overthrew a dictator).

The United States Institute for Peace – suggestions for non-violent action: https://www.usip.org/issue-areas/nonviolent-action

Adbusters for inspiration on using creative advertising to get the message across. Also look for the Adbusters Field Guide to Virtual Warfare.

https://www.ende-gelaende.org/en/ an anti-coal system change movement in Germany. The website is also in English.

The Wild Lily Student Movement, Wild Strawberry Movement and Sunflower Movement in Taiwan, which illustrate what can be achieved peacefully.

Clearing the PR Pollution that Clouds Climate Science – www.desmogblog.com

The "System change not climate change" movement. The US arm was co-founded by Richard Smith, the author of the excellent book, *Green Capitalism, the God that Failed* https://systemchangenotclimatechange.org/

https://www.attac.org/ – opposing neoliberal globalization, with a focus on fighting for greater regulation of the finance industry.

Endnotes

1. Zoonotic host diversity increases in human-dominated ecosystems, Nature, August 2020.
2. The Global Plastic Calamity, Bluewater, February 2019.
3. How air pollution is destroying our health, World Health Organisation, May 2018.
4. If pre-covid-19 emissions continue as before it will take 14.8 years. Pre Covid-19 emissions were growing by 2% a year. This leads to a tipping point in 12 years. Post Covid-19, emissions are expected to fall by 5.5% in 2020. If they stay at that level, the tipping point will be in late 2035.
5. Kevin Anderson, Deputy Director, Tyndall Centre for Climate Change Research, 2009. Prof. Anderson considers that "a 4°C future [relative to pre-industrial levels] is incompatible with an organised global community, is likely to be beyond 'adaptation.'" "If you have got a population of nine billion by 2050, and you hit 4°C, 5°C or 6°C, you might have half a billion people surviving." (Fyall 2009). See Disaster Alley Report, Dunlop, and Spratt July 2017.
6. Dunlop and Spratt, *What Lies Beneath: The scientific understatement of climate risks*, 2017, Rockstrom et al, *A roadmap for rapid decarbonization*, Science, 2017, and Dunlop, Disaster Alley: Climate change, conflict and risk, 2017.
7. UNEP's annual *Emissions Gap Report*, 2019. https://www.unenvironment.org/news-and-stories/press-release/cut-global-emissions-76-percent-every-year-next-decade-meet-15degc
8. Gross Domestic Product
9. OECD. (October 2014). *How Was Life? Global well-being since 1820.*
10. Smith, Adam, *An Inquiry into the Nature and Causes of the Wealth of Nations*, p. 148

11. Ibid. Chapter XI, Part III, Conclusion of the Chapter, p. 292
12. Smith, Richard, 2014 *Green Capitalism, the God that Failed.*
13. See for example "The equal pay revolutionaries", Financial Times, 8–9 August 2020.
14. *Conflict And Violence In The 21st Century: Current Trends As Observed In Empirical Research And Statistics*; World Bank Group, 2016.
15. Letter to the Egyptian Gazette, August 25, 1964.
16. https://taz.de/Oekonomin-ueber-Meinungsmanipulation/!5585707/
17. Lu Chien-Yi, *Surviving Democracy – Mitigating Climate Change in a Neoliberalised World*, Routledge 2020, Introduction, p. 2
18. Scheidler, Fabian, *The End of the Megamachine*, Zero Books, 2020.
19. Sweden had 567.9 Covid-19 deaths per million, higher than the US, at 518.7 per million, and just behind Italy, Spain, and the United Kingdom. Coronavirus (Covid-19) deaths worldwide per one million population as of August 17, 2020, by country; statista.com
20. Exxon's Climate Denial History: A Timeline | A review of Exxon's knowledge and subsequent denial of climate change; Greenpeace.
21. Big Oil's Real Agenda on Climate Change, March 2019. InfluenceMap.
22. https://en.wikipedia.org/wiki/All_your_base_are_belong_to_us
23. https://www.nytimes.com/interactive/2019/10/29/climate/coastal-cities-underwater.html
24. The Militant; February 17, 2003; Volume 67, No. 6
25. Based on the 1944 movie Gaslight, a man manipulates his wife until she believes she is losing her mind.
26. A neuro-degenerative fatal brain disease of cattle, which exhibited in humans as Creutzfeldt-Jakob disease (CJD).
27. Confusingly, insect numbers are falling (https://www.

theguardian.com/environment/2019/feb/10/plummeting-insect-numbers-threaten-collapse-of-nature) while insect-borne diseases are increasing (https://www.cdc.gov/vitalsigns/vector-borne/index.html)

28. https://www.ghsindex.org/about/
29. UNEP's annual Emissions Gap Report, 2019. https://www.unenvironment.org/news-and-stories/press-release/cut-global-emissions-76-percent-every-year-next-decade-meet-15degc

TRANSFORMATION

Transform your life, transform your world – Changemakers Books publishes for individuals committed to transforming their lives and transforming the world. Our readers seek to become
positive, powerful agents of change. Changemakers Books inform, inspire, and provide practical wisdom and skills to empower us to write the next chapter of humanity's future.
www.changemakers-books.com

The *Resilience* Series

The Resilience Series is a collaborative effort by the authors of Changemakers Books in response to the 2020 coronavirus epidemic. Each concise volume offers expert advice and practical exercises for mastering specific skills and abilities. Our intention is that by strengthening your resilience, you can better survive and even thrive in a time of crisis.
www.resilience-books.com

Adapt and Plan for the New Abnormal - in the COVID-19 Coronavirus Pandemic
Gleb Tsipursky

Aging with Vision, Hope and Courage in a Time of Crisis
John C. Robinson

Connecting With Nature in a Time of Crisis
Melanie Choukas-Bradley

Going Within in a Time of Crisis
P. T. Mistlberger

Grow Stronger in a Time of Crisis
Linda Ferguson

Handling Anxiety in a Time of Crisis
George Hoffman

Navigating Loss in a Time of Crisis
Jules De Vitto

The Life-Saving Skill of Story
Michelle Auerbach

Virtual Teams - Holding the Center When You Can't Meet Face-to-Face
Carlos Valdes-Dapena

Virtually Speaking - Communicating at a Distance
Tim Ward and Teresa Erickson

Current Bestsellers from Changemakers Books

Pro Truth

A Practical Plan for Putting Truth Back into Politics
Gleb Tsipursky and Tim Ward

How can we turn back the tide of post-truth politics, fake news, and misinformation that is damaging our democracy? In the lead up to the 2020 US Presidential Election, Pro Truth provides the answers.

An Antidote to Violence

Evaluating the Evidence
Barry Spivack and Patricia Anne Saunders

It's widely accepted that Transcendental Meditation can create peace for the individual, but can it create peace in society as a whole? And if it can, what could possibly be the mechanism?

Finding Solace at Theodore Roosevelt Island

Melanie Choukas-Bradley

A woman seeks solace on an urban island paradise in Washington D.C. through 2016-17, and the shock of the Trump election.

the bottom

a theopoetic of the streets
Charles Lattimore Howard

An exploration of homelessness fusing theology, jazz-verse and intimate storytelling into a challenging, raw and beautiful tale.

The Soul of Activism

A Spirituality for Social Change

Shmuly Yanklowitz

A unique examination of the power of interfaith spirituality to fuel the fires of progressive activism.

Future Consciousness

The Path to Purposeful Evolution

Thomas Lombardo

An empowering evolutionary vision of wisdom and the human mind to guide us in creating a positive future.

Preparing for a World that Doesn't Exist - Yet

Rick Smyre and Neil Richardson

This book is about an emerging Second Enlightenment and the capacities you will need to achieve success in this new, fast-evolving world.